Thinking About Technology

Foundations of the Philosophy of Technology

Joseph C. Pitt
Virginia Polytechnic Institute and State University

Seven Bridges Press, LLC
135 Fifth Avenue, New York NY 10010

Publisher: Clayton Glad
Managing Editor: Katharine Miller
Project Manager: Electronic Publishing Services Inc., N.Y.C.
Production Services and Page Makeup: Electronic Publishing Services Inc., N.Y.C.
Cover Design: Andrea Barash Design
Printing and Binding: Versa Press, Inc.

Library of Congress Cataloging-in-Publication Data

Pitt, Joseph C.
Thinking about technology: foundations of the philosophy of technology/
Joseph Pitt. — 1st ed.
p. cm.
Includes bibliographical references (p.).
ISBN 1-889119-12-1
I. Technology—Philosophy. II. Title.
T14.P55 2000
601—dc21

99-42441
CIP

Manufactured in the United States of America
10 9 8 7 6 5 4 3 2 1

For Donna

Contents

Preface ix

1. Looking for Definition: Epistemology and Technology 1
The Limits of the Philosophy of Science 7
The Social Dimensions of Science 8
Defining "Technology" 9

2. Modeling Humanity at Work 13
Practical Reason and Rationality 16
Rationality 19
Conclusions 23

3. Locating the Philosophy in Technology 25
Counterpart Questions 26
Technological Knowledge: Setting the Stage 28
Scientific Knowledge 32
Engineering Knowledge 35
Philosophical Problems 38

4. Technological Explanation 41
Scientific Explanation 42
Technological Explanation 43
Technical Explanation 45
An Example: The Case of the Hubble Space Telescope 51

5. Technology and Ideology 66
Some Preliminaries: About Heidegger 67
Technology and Ideology 70
Technology and Values 82

6. The Autonomy of Technology 87
Trivial Autonomy 88
The Process of Technology 90
Common Sense 91
Galileo and the Telescope 93
Geometry as a Technology 96
Technology and the Dynamics of Change: Autonomy Socialized 97

7. Technology, Democracy, and Change 100
Innovation and Control 100
The Race between Technology and Morality 112
Conflicts of Values and Technological Change 113
Changing the Way Humanity Works 114

8. Scientific Change and The Technological Infrastructure of Science 122
Theories of Scientific Change 122
The Technological Infrastructure of Science 125
Conclusions 138

Bibliography 139
Index 143

Preface

IT ALMOST GOES WITHOUT SAYING that technology is a pervasive feature of contemporary culture. In this book I will be arguing, among other things, that technology is more than that; it is a defining feature of the human condition. It is, therefore, the job of philosophy, the form of inquiry best suited to focusing our thinking about the "big" questions, to make sense of technology and to help us understand its workings as well as its impacts on our lives and values.

Unfortunately, the kind of attention that contemporary philosophers usually give to matters of technology consists primarily of social criticism. Post-World War II philosophical treatments of technology have been primarily negative, taking the form of critical denunciations of the negative effects of technology on human values and on human life. On the whole, social criticism is a perfectly respectable aspect of philosophical work. Most social criticism of technology comes in one of three forms: (a) critiques based primarily on ideological considerations, e.g., Marxists or Earth-Firsters; (b) critiques based on projections of the consequences of not rejecting a technology or developing a new technological project, e.g., huge dams or nuclear reactors; (c) a combination of (a) and (b). In what follows I will have something to say about each form of social criticism. Briefly, with respect to (a), I argue that ideological critiques contribute little to rational consideration of the merits or demerits of a technology. With respect to (b), I am most concerned to argue that we need to assess our epistemological assumptions (about, for example, the reliability of forecasting techniques used to predict the consequences of specific technological innovations or changes) before we go on to use various claims generated in the context of those assumptions as the basis for a critical assessment of the merits of the technology in question. On a more fundamental level, the level with which I am most concerned, we need to know how to evaluate, for example, a technological explanation, before we go on to use that explanation as the basis for a critical assessment of the merits of the technology in question. Knowing how to evaluate an explanation is essentially an epistemological (and perhaps a logical) problem (depending on the type of explanation).

This book presents a case for the logical priority of epistemological issues over social criticism in the "order of knowing." Understanding what we know about technology, and understanding *how we know that what we know is reliable*, are the prerequisites to offering sound evaluation of the effects of technologies and technological innovations on our world and on our lives.

There is another reason for insisting on the priority of epistemological questions in philosophical discussions of technology. This has to do with the nature of philosophy as an activity in itself. Philosophy aims at explaining how everything "hangs together," at showing how values and hopes for the way things should be either affect, ignore, improve on, or debilitate the way things are. Philosophy tries to explain the world by explaining the interrelatedness of all its parts, both the way it is and the way we would like it to be. It is important, therefore, that philosophical explanations assist this process. As it happens, the social critics of technology do not provide analyses that can be used in this way, for they begin by assuming a privileged point of view about the way things ought to be, rendering their conclusions and their analyses incapable of handling such not-so-trivial questions as: How do you know you are right? Whatever else it is, philosophy must be self-critical. It has no one single methodology, but even so, whatever methodology is employed must be capable of reflexivity.

The word "philosophy" in its original Greek meant love of wisdom. It does not follow now (nor did it for the Greeks) that one who practiced or practices philosophy is wise, nor does it follow that the result of philosophizing is wisdom. Loving wisdom does not necessarily make you wise, just as loving beauty does not make you beautiful. To do philosophy is be engaged in the *search* for wisdom. Considered this way, philosophy is not so much a body of truths as a *process* or a means of exploring. The history of philosophy is the history of that search, and it can be seen as having the form of a dialogue. It is a discussion with participants drawn from the present and the past on issues of long-term human concern.

Viewing the history of philosophy as a dialogue has its advantages and disadvantages. One advantage is that the concerns of philosophers can best be made sense of when viewed in context. On the other hand, understanding philosophy as a dialogue can be frustrating. The frustration comes not only from the amount of time it takes to complete some aspect of the dialogue, but also because the questions change over time. And while discussions of the general issues range over epochs, it is also true that things change over time. In philosophy, the importance, meaning, and viable solutions to these long-lasting issues also change as the culture changes. The result is that philosophy is a *continuing* and *dynamic* dialogue.

Well then, you may ask, what is the difference between philosophy and any other discussion? The difference has to do with the aim of philosophy. Perhaps the most eloquent and provocative account of the aim of philosophy was put forth by Wilfrid Sellars. As we will see, Sellars sets his sights rather high. From a philosophical point of view, however, there is nothing wrong with aiming high. In fact, settling for less than the most we can conceive of is what should be avoided. Philosophy, then, in the words of Wilfrid Sellars, is concerned

> to understand how things in the broadest sense of the term hang together in the broadest sense of the term. Under 'things in the broadest possible sense' I include such radically different items as not only 'cabbages and kings,' but numbers and duties, possibilities and finger snaps, aesthetic experience and death. To achieve success in philosophy would be, to use a contemporary turn of phrase, to 'know one's way around' with respect to all these things, not in that unreflective way in which the centipede of the story knew its way around before it faced the question, 'how do I walk?' but in that reflective way which means that no intellectual holds are barred. (Sellers 1963, 1)

It should be obvious why I think Sellars is aiming too high. Except perhaps for Aristotle, up to this point no one has managed to make complete sense of the way things are. This is not to say that no one has tried. Quite the contrary, the history of philosophy is the history of our efforts to achieve that very goal. The fact remains, however, that no completely satisfactory philosophical system has been produced. That is, we have yet to produce a system that meets Sellars' conditions. There are many reasons for this failure, some of which we will consider below. But, returning briefly to the role of questions about technology in the philosophical dialogue, to use the lack of a satisfactory philosophical system as an explanation of the failure of philosophers to come to grips with technology takes the cheap way out; on those grounds, since no one has succeeded in making sense of everything taken together, nothing has been explained at all.

But, as we are well aware, it is simply not true that nothing has been explained. For example, we find that scientific inquiry continues to help us make increasing sense of the structure of the universe and our place in it as physical beings. Historical and cultural studies continue to examine and probe the significance of humanity and its works in and over time. And while no complete picture has emerged that truly accomplishes what Sellars set as the goal of philosophy, it just may be that this is due to the incompleteness of our understanding. Some of the things we have yet to appreciate fully are the complexity of humanity's own activity, the ways in which we respond to the environment that we are continually changing, and the effects of our impact on the environment on us. That, plus the complexity of the universe, makes a completely systematic and comprehensive explanation of "life, the universe, and everything" utopian.[1] That our goal is utopian does not mean that it should be abandoned and that we should give up trying to explain the universe and our place in it. While it might be true that we should abandon some goals, such as trying to fly without physical assistance, it does not follow that all goals beyond our immediate reach should be abandoned, even if they are in principle beyond our reach. Specifically, goals such as seeing how everything fits together should not be rejected, even if the possibility of their being attained is remote. We need to retain such utopian goals because they help guide our activity in situations that transcend individual aspirations. Examples of such goals are to be found in the sorts of programs human beings support even when they are not of particular relevance to their own discrete lives, such as welfare for the needy when one is affluent.

Sellars' conception of the goal of philosophy seems to be one of these utopian goals we should not abandon, even though we can never really complete our picture of how everything in the world hangs together. Philosophy, if it is truly concerned with understanding how to make our way around in the world, is necessarily and continually changing. As science tells us more and more, learning our way around that world must be an enterprise in constant need of updating, and so the philosophical questions we have been working on need constant rethinking. But to admit that no complete philosophy is possible is not to give up doing philosophy as a search for wisdom. In fact, philosophy remains alive and becomes increasingly important as the world becomes increasingly complex. The need to bring some degree of understanding to the apparent chaos of contemporary civilization becomes increasingly important as the world of human possibilities becomes more complex.

Central to my concerns is the disturbing tendency of the social critics and others to speak about "Technology" as if it were one thing. Try as I may, I cannot find the one thing. I can find automobiles, power stations, even specific government offices, but nowhere can I locate *Technology* pure and simple. It is the same problem I have when I try to find Science. I *can*

---

1. The expression "life, the universe, and everything" is borrowed from Douglas Adams's book by that title, Volume Three of his five-volume *Hitchhiker's Guide to the Galaxy* trilogy, wherein the futility of comprehensive understanding is extolled. (The series was announced as a trilogy, and is still so labeled, but it has five volumes, indicating perhaps an extension of the author's philosophical position.)

locate physicists, biologists, journals, laboratories, etc. But I can't find the thing called "Science." And so, in this essay one of the themes, which I discuss in a number of different ways, is that there is no one single thing called "Technology." In this respect, the definition of "Technology" I offer—humanity at work—should be seen as punctuating the need to stop talking about *Technology* simpliciter and to start focusing on the specific problems we encounter and the techniques, materials, etc., we employ, as well as the consequences of using these techniques and materials to solve those problems. And while seeking to show how things hang together, I also work toward a conclusion that separates two concepts that currently appear to be wedded in our understanding of the world but should not be. These are the notions of "Science" and "Technology."

On the surface this sounds somewhat paradoxical. On the one hand, I am saying that in our complex world we need philosophical analyses to help us make sense of what we are doing. On the other hand, I am arguing that we should abandon, at least in this one area, efforts to oversimplify by talking about technology. How can we make sense of things by making the world more complex? The answer is that in a misdirected search for simplicity we end up misconstruing the actual situation, thereby opening ourselves to inappropriate actions. We should not confuse understanding with simplicity. The world is a very complex place, and we should not deceive ourselves into thinking otherwise, lest we suffer the consequences.

Finally, given the prevailing intellectual fad of hostility to anything concerned with establishing conceptual priority, something should be said about how it is possible in the current climate even to consider proposing a work on the foundations of the philosophy of technology. Recently it has been popular to dismiss the search for the intellectual foundations of inquiry.[2] Despite its popularity, it doesn't follow that that conclusion is warranted. The objection to claims about foundations comes from a slippery-slope argument, which acknowledges that our knowledge of the world is underdetermined by the evidence. Since scientific knowledge is said to be the foundation of all we know about the world, and since that knowledge is undermined by that underdetermination, science, it is concluded, is no better than any other form of inquiry. That is, whatever it is we come to know scientifically about the world, it is not enough to demonstrate conclusively that that is the way the world must be. Hence, it allegedly follows, science is just one form of inquiry among others, with

---

2. Consider Michel Foucault and Jacques Derrida. In Foucault's *The Order of Things* (1966), he attempts to undermine the foundations of human sciences by exposing the contingencies upon which they rest. In the essay "Structure, Sign and Play in the Discourse of the Human Sciences" (1967), Derrida attempts to argue against the very possibility of a center or foundation.

no obvious claim to superiority. Thus, it is concluded, since science is supposed to be our most secure form of knowledge and since it cannot lay claim to absolute certainty, then there is no form of knowledge with sufficiently firm foundations to warrant claims of superiority to any other, and, so the argument continues, if science can't make the claim, then nothing else can either.

The fact that scientific knowledge is underdetermined by the evidence does not warrant the conclusion that no form of knowledge can be superior to any other. To my mind the best accounts of knowledge allow that knowledge is defeasible. What we know now may change in light of new discoveries. From this is does not follow that we know nothing now. Surely, to some extent, the resolution of this argument will be a function of how we define "knowledge." And I for one have a jaundiced view of efforts to find universal and eternal definitions of anything. For surely, whatever claims we make, whatever definitions we form, should be revisable in the light of continuing human experience. With that assumption in hand, I will opt for an operational, call it a pragmatic, approach to knowledge. This means that I will look for the hallmark of knowledge to be successful action. If, on the belief that x causes y, when I do x, y happens consistently, that is good enough for me. With appropriate refinements, this account should satisfy most skeptics as a working definition. That being the case, I think it fair to say that science is the most successful of human activities at producing reliable knowledge. It therefore has and ought to have a place of epistemic privilege, i.e., scientific knowledge ought to be preferred to other kinds of knowledge. And if that makes it foundational, so much the better.

Essentially I am suggesting that to seek for the foundations of intellectual inquiry is not necessarily to be committed to the belief that it is possible to find universal truths. Instead of true propositions, I wish to emphasize successful methods. It is in this sense that this work is a search for the foundation of the philosophy of technology. I am looking for the best methods to employ to understand both the varieties of technologies with which we are presented and the issues they present to us in a way that makes it possible to see how it all hangs together.

The structure of the book is fairly straightforward. First, I develop a framework for thinking about specific issues that arise in the context formed by a specific technology. Second, I introduce and explore a set of concepts that are counterparts to concepts that have already been the object of intense analysis by philosophers of science. The strategy here is straightforward. Philosophers of science have examined in detail a number of concepts integral to our understanding of what makes science what it is. If science and technology are as closely linked as assumed, there ought to counterparts to these concepts that apply to technology. Thus, following up the concept of a scientific explanation, I look at the conditions for a technological explanation. It turns out that they are very different from the Standard View of scientific explanation. On the basis of a similar examination of several such counterpart concepts, I suggest that maybe science and technology ought not to be thought of as so closely linked in our thinking as we currently think they are. The bottom line is this: philosophical questions about technology are first and foremost questions about what we can know about a specific technology and its effects and in what that knowledge consists. This amounts to knowing what we as human beings can know about the world and our impact on it. That is why I think epistemological issues should be addressed before we engage in social criticism. I then proceed to attack a set of assumptions about "technology" put forth by social critics. Whatever else "it" may be, I argue that technology is not autonomous or a threat to democracy. I further argue that talking about technology in this way misleads in important ways. Finally, I address the problem of technological change. After examining extant models of scientific change, showing them to be inadequate, I explain the inadequacy by appeal to their failure to take into account the technological infrastructure of science and the manner in which science is embedded in and fundamentally tied to it.

Some of the material contained in this book has appeared in part in separate articles. I wish to thank the publishers of *Philosophy of Science*, *Philosophy & Technology*, *The Electric Journal of the Society for Philosophy and Technology*, *SPT Press*, *The Center for Phenomenological Research*, and *Wall Press*, for the use of this material.

It is impossible to thank everyone to whom I am indebted for this project. It would, however, be difficult indeed not to acknowledge Karen Synder, Richard Burian, Harlan Miller, Ann LaBerge, Morton Tavel, Marjorie Grene, Paul Durbin, Carl Mitchem, Peter Kröes, Gary Hardcastle, Richard Hirsch, and Andrew Garnar for their assistance and critical comments. My father, Lewis Pitt, kept me supplied with boundless articles and resources. A very special thank you goes to my students who over the years have exhibited a never-ending and cheerful willingness to tell me how wrong I am. My editor, Clay Glad, is a brave man. Donna sustains me.

CHAPTER ONE

Looking for Definition: Epistemology and Technology

I WANT TO DISCUSS TECHNOLOGY in a such a manner as to allow those discussions to be informed by and, in turn to inform, the rest of our philosophical as well as our daily worries. It would help matters if there was a generally accepted definition of "technology." Many definitions exist, but there is little agreement on which one is the best. I propose, therefore, to start from scratch.

One way to begin is to distinguish between technology and other things with which it has been closely related. The activity with which technology has been most closely paired recently has been science. Unfortunately, most of what has been said in the abstract about the relation between science and technology has not been very helpful. In large part this is due to a series of assumptions about the nature of knowledge, in particular scientific knowledge. Science is a knowledge-producing enterprise. Much work in the philosophy of science has been epistemological, i.e., it has been concerned with the nature of scientific knowledge, its justification, structure, and relation to certain metaphysical issues. So, if a definition of technology is to be found by distinguishing technology from science, I suggest we look at the epistemological dimension of technology *in its own terms* and not as necessarily tied to science. To assume some crucial relation between science and technology begs the question.

Technology does have an important epistemological aspect to its character and, furthermore, as I have suggested, it is this epistemological dimension that is crucial to the philosophical placement of technology-related issues in the philosophical conversation. However, the standard account of epistemological issues has been formulated in such a way as to misdirect our approach to an understanding of the relation between science and technology. In particular, there are three mistaken assumptions about the epistemological relations between science and technology that have governed much of our thinking about these matters. These assumptions are:

(1) a distinction between theoretical and applied knowledge, with science represented on the side of theoretical ("pure") knowledge;

(2) a hierarchical account of knowledge, with "pure" scientific knowledge presented as superior to applied knowledge;

(3) characterizing technology as applied knowledge, hence inferior to science.[1]

The first of these, the venerable distinction between "pure" and "applied" with respect to knowledge and science and technology, has it that science is pure and technology is applied. However, it is very difficult to determine what is supposed to be pure or applied in either area. If the proposed answer is "knowledge," then the claim becomes "science is pure knowledge and technology is applied knowledge," and this is surely false, since, as we shall see, science is not pure knowledge. Likewise, if technology is supposed to be applied scientific knowledge, this view must be rejected, for many technologies do not necessarily require prior grounding in the theoretical work of science.

Without trying to define "knowledge," we can, nevertheless, agree that the product of science is knowledge. To see this requires invoking (rather than attacking) a different distinction, this time between the *process* whereby we produce knowledge and the *product* of that process.

Science is a process composed of a large number of diverse activities undertaken by a variety of individuals in various settings, mostly in institutions such as universities and major laboratories established in governments or by business. If processes can be said to have goals, then our best way to understand the scientific process is normatively to characterize its goal as the production of knowledge. To accomplish this goal, scientists use theories, among other things such as instruments. Scientific theories consist of sets of assumptions about the population, structure, and behavior of a domain of physical entities whose existence is postulated in order to explain the behavior of objects we encounter or observe in everyday

1. This last assumption is such a popular claim that in some senses it is hard to document. It does form a background, however, to arguments about funding basic research, as well as providing the basis for distinguishing within science between different forms of activity by scientists, e.g., experimentalists versus theoretical physicists, field versus lab biologists, etc.

life. For example, the atomic theory of matter can be used to explain why some tables are hard while water at 50°F is not. Part of the process of scientific inquiry involves testing the assumptions of such theories, revising them in the light of new information, extending them into new domains, testing further, and attempting to find the limits to which these theories can be pushed before they have to be abandoned and replaced by new theories, new assumptions, etc. Because the scientific process *must* use theories to guide its research, the research is already infused with the assumptions of the theories and the methods being used to explore its domains. And when we change theories we do not escape assumptions; we simply replace one set of biases with another. So, in this very fundamental sense, science cannot be pure, for it is constrained and directed by its theories because the theories incorporate assumptions and preconceptions about the methods to use and the domains under investigation.

But there is another sense of "pure" that is relevant here. Science might be thought to be pure in the sense that its goal is "knowledge for its own sake." That is, the product of scientific inquiry is knowledge, but that knowledge is not the result of someone saying "we need to know *x*, so we can do *y*." When it is said that scientific knowledge is pure knowledge, or knowledge for its own sake, what is meant is that there is no reason for obtaining that knowledge *other than the goal of knowing itself*. This is often invoked as the justification for funding basic scientific research.

Unfortunately, this conception of science as the pursuit of knowledge for its own sake itself rests on some questionable assumptions. In fact there is a certain incompatibility between what we accept as knowledge and the idea of knowledge for its own sake. That incompatibility stems from the fact that the production of knowledge is, as we shall see, a community enterprise. The process/product distinction used above can also be of assistance here. But before we bring it into the picture we need to do a bit of history of philosophy.

The theory of knowledge has been an essential component of the philosophical landscape since philosophy became an important aspect of our culture. Enduring theories of knowledge date back to Plato and Aristotle. In the seventeenth century, with the rise of the New Science, answers to the philosophical questions about the nature of knowledge were attempted by individuals committed to the idea that science required a solid epistemological base. There were two major schools of thought at this point: empiricism and rationalism. Their champions included John Locke, George Berkeley, and David Hume for the empiricists, and Rene Descartes, Baruch Spinoza, and Gottfried Leibniz for the rationalists. There were significant differences between these two camps on a number of issues, but common to both was an emphasis on the means whereby an *individual* acquired the basic material from which he formulated his beliefs and the process by which he transformed that material into knowledge. This total focus on the role of the individual embodied a crucial confusion between the process by which individuals acquired *beliefs* and the resultant production of *knowledge*. For instance, it was assumed that if the beliefs were acquired in a certain way, i.e., through experience, or if they could be traced to a certain kind of source, one that guaranteed certainty, then the content of the belief could be construed as knowledge. This view of knowledge is retained today by adherents of the view that knowledge is justified true belief. It is a view I wish to reject in favor of a more complex account in which the process by which individuals acquire and justify beliefs is but the first stage in the search for new discoveries.

There is no difficulty with the idea that individuals conduct inquiries, discover new things, and offer up candidates for inclusion into the body of accepted and integrated claims we recognize as knowledge. Let us refer to such claims as *candidate-claims*. Trouble begins when it is also assumed that the individual producing the candidate-claim is solely responsible for determining whether or not that candidate-claim counts as knowledge. To make this assumption is to take the path of traditional empiricist and rationalist epistemologies, with all their attendant difficulties. The way out of those difficulties is to make the *community* and not the individual the determiner of what counts as knowledge, that is, the ultimate status of a knowledge candidate-claim is determined by the community and not the individual. This approach is the result of the insight of the American philosopher Charles Saunders Peirce (1839–1914). It represents a major break with traditional epistemology and constitutes one of the key assumptions of the philosophical school of thought Peirce initiated that we know today as *pragmatism*.[2] It might be surprising to some to find out that this social component to knowledge comes from a thoroughgoing philosophical position and not from the less sophisticated views of contemporary thinkers committed to what has come to be called social constructivism.[3] Further, Peirce's view on the role of the community of observers is also

---

2. Although the philosophical stance of the author was not announced as "pragmatism," Peirce's 1868 essay, "Some Consequences of Four Incapacities" (CP: 5.264-317) outlines many of the concerns for this movement. Also see "The Fixation of Belief" (CP: 5.358-387) and "How to Make Our Ideas Clear" (CP: 5.388-410), esp. paragraphs 405 to 410 of the latter, for Peirce's account of the social character of science.

3. For examples see Collins 1981; Collins 1985; Collins 1990; Bijker et al. 1987.

tied to a deep-seated form of scientific realism, thereby avoiding a major critique of constructivism—its relativism. I do not share Peirce's realism, but I also reject the relativism of the Constructivists. In Chapter 8 I develop an approach that takes both the social component of knowledge and the methodological insights of realism seriously while rejecting Peirce's version of convergent realism.

Pragmatism also endorses the view that if the community is the arbiter of knowledge, then *successful action* is its criterion. That is, the ultimate test of what is to count as knowledge is determined by our ability to act successfully on that knowledge.[4] If the world is reported to be a certain way, the final test of that candidate-claim will be the success of an individual or a group acting as if the world were in fact that way. Not only is knowledge determined by the limits of action in this fashion; its purpose is action. Thus we seek to discover or uncover the way the world is in order to make our way around in it better.

It well may be the case that certain individuals involved in the process called science do whatever they do simply because of a certain innocent curiosity or because they derive personal delight from these activities. But we should not confuse the collective of activities called science with the personal delight of any one person. Whatever candidate-claims an individual proposes, it is the conclusions of the community of investigators that determines whether or not that candidate-claim is to count as knowledge at this time. That is one reason why publishing is so important in science. Publication amounts to community endorsement of those findings, i.e., candidate-claims. But the grounds for acceptance for publication have to do, for the most part, with how the candidate-claims fit in with what has already been accepted, where these already accepted claims have been shown to have real-world use, even if only in terms of offering explanations of the way the world works. It would seem, then, that if we reject the old-fashioned epistemologies of the traditional empiricists and rationalists, we must also reject the idea that it makes sense to talk about pure science producing knowledge for its own sake.

If we now turn to the other side of the coin and consider the idea that technology is merely applied science and, hence, applied knowledge, we don't fare much better. To begin with, there are technological items that have been constructed and used without a theoretical knowledge base to explain how they work. Consider the roads of Rome and Galileo's telescope. The Romans did not rely on "scientific" principles to guide the construction of their roads or aqueducts or catapults. When Galileo built his telescope there was no scientific explanation available to account for

4. This notion is first developed in "The Fixation of Belief," esp. CP: 5.384-387.

why the telescope did what it did.[5] In other words, these technologies themselves represent knowledge insofar as they are the useful results of successful actions. They are not the result of the application of some abstract set of theoretical principles. Likewise for gunpowder, boats, water mills, and so on.

Second, the history of science shows clearly why facile accounts of the relation between science and technology fail: *the historical relationship between science and technology is not straightforward and generalizable.* This means, minimally, that the relations between science and technology are vastly more complicated than the pure/applied distinction would suggest. If anything, *the relation between science and technology is symbiotic and mutually nurturing, with theory and mechanisms feeding on and fueling each other under the influence of and in response to a variety of cognitive, social, and economic pressures.*

Given its complexity and its importance, the analysis of this relationship would be exactly the sort of problem that should attract the attention of historians and philosophers of history and technology. Historians both of history and of technology *have* addressed some of these issues by providing detailed case studies of specific episodes. Contemporary philosophers of technology have, for the most part, however, viewed their role as that of social critic, and they have concentrated on the impact of various technologies on values and on the environment,[6] leaving the question of the epistemological relations between theory and mechanism largely unexplored.[7] And philosophers of science have been unwilling to consider the role and the impact of technology on science for at least two different reasons. The first involves the set of epistemological assumptions we are currently examining and the second is the self-limiting manner in which issues in the philosophy of science have been characterized, especially in this century, and in the context of raising the social status of Western science.

---

5. After the fact, Galileo constructed an explanation for why his telescope worked. It was not a very good explanation, since it rested on a false optical theory. See Drake 1978, 131-38.

6. Clearly, Langdon Winner's work exemplifies this sort of social criticism (see Winner 1977; Winner 1986). Also consider Jacques Ellul, who is concerned with the increasing rationalization of society through technology. In a more sophisticated manner, Heidegger engages similar issues. Another example of the prevalence of such social criticism is the latest edition of Albert Teich's anthology, *Technology and the Future*. The majority of these articles analyze how a given technology changes values, society, or the environment. Little, if any, consideration is given to the epistemic and ontological dimensions of technology.

7. There are some exceptions, specifically Bachelard 1934; Hacking 1983; Galison 1987; Galison 1997.

Section 1. The Limits of the Philosophy of Science

The dominant school of thought in the philosophy of science in the West in the twentieth century has been Logical Positivism and its varieties. This is not to say its program has been completely successful. It is rather to acknowledge the fact that much of what the positivists said about the fundamental character of science has had tremendous influence on the way both philosophers and nonphilosophers have thought and continue to think about the epistemic authority of science.

Positivistic philosophers of science have been concerned with explicating the concept of a scientific theory, the nature of scientific explanation, and other specific issues such as the problem of confirmation. Their approach to these problems was to worry about the *logic* of these concepts and relations. This essentially meant that the actual theories, explanations, etc., used by scientists were to be ignored in favor of an examination of the abstract consequences one was committed to if these notions were successfully defined, where "success" was to be measured by logical structure and universality of application. Since the objective in such investigations was the justification of the authority of scientific results in general, the peculiarities of particular scientific episodes were perceived as having little relevance.

Although current research in the philosophy of science now extends beyond the narrow positivist agenda, only recently has the range of issues been expanded to include questions such as the changing character of science itself, since one of the consequences of the positivist's view was that the essential nature of science didn't change. Until this expansion, there was little apparent need for philosophers to worry about the details of the technological mechanisms used in various experiments or the social, political, and economic considerations surrounding the ongoing process of scientific activity, since these matters were seen as having little to do with reconstructing the logic of the concepts in general. In other words, positivistically influenced philosophers were concerned to develop a broad and general account of the grounds of scientific knowledge, an account that relied heavily on the authority of logic. But the fact of the matter is that the details of daily scientific work result as much from the psychology of individual scientists, institutional settings, and the social dynamics of group interaction as from any other factor.[8] Positivists, on the other hand, argued that the authority of scientific knowledge cannot be a function of the psychology of group dynamics—it must rely on more basic principles of good reasoning. But in concentrating on the logic of

8. Galison illustrates this in both *How Experiments End* (1987) and *Image and Logic* (1997).

concepts and ignoring the actual activities of scientists, philosophers left themselves open to the legitimate charge of being irrelevant. What good does a philosophical account of science do us if it bears no relation to the actual activity of science?

When notice was taken of factors beyond the logic of concepts, philosophers of science were simultaneously provided with the opportunity to integrate the philosophical dimensions of technology into the general dialogue, but they did not take advantage of it. Instead it was left to an historian of science to force the issue. In *The Structure of Scientific Revolutions*, Thomas Kuhn provided a powerful argument philosophers had to acknowledge, in which factors other than those of pure reason or logic were shown to play a serious role in determining the options available to scientists and the avenues of research they develop. Kuhn showed that science does not just roll on unimpeded toward the truth propelled purely by logic.

An examination of the kinds of social, political, and economic factors Kuhn introduced becomes increasingly important when we try to understand the structure and development of contemporary science. Furthermore, Kuhn's theory of the development of science relativized particular achievements in science to discrete combinations of theories and procedures he called *paradigms*. While Kuhn did not explicitly speak of the contributions of technological innovations to the development of paradigms, his approach opened the door to such considerations. Subsequent case studies of discrete episodes in the history of science have revealed the fascinating way in which the tools and experimental devices scientists use to test out their theoretical ideas influence the development of the theories, even to the point of contributing to their rejection, e.g., Michelson-Morley.[9]

Section 2. The Social Dimensions of Science

Many features crucial to the past development of science were a function of technological considerations that are also, to no one's surprise, turning out to be crucial today. Not only are specific instrumentalities involved, as in the past where advances in our understanding had to wait for the development of items such as the telescope, but today another type of "mechanism" is intimately involved in the doing of modern science in a way that critically determines what can and cannot be done. This is the bureaucracy of the institution of science. It includes government funding agencies, peer review, career competition, journals, and the educational process. In the context of the *doing* of science, many of these factors can

9. C.f. Galison 1997 for a nice examination of some of the factors involved here.

be viewed as tools to be used and manipulated to achieve specific ends, i.e., as technologies.

That the infrastructure of the scientific establishment can be viewed as a set of tools should come as no surprise. Although we often tend to think of tools in more simpleminded terms, there is nothing in principle to keep us from recognizing more complicated items that are as much tools as are hammers and wheels.[10] But, it might be argued, aren't we risking confusion if we reduce all of humanity's achievements to mere tools? Don't we really need finer-grained analyses, ways of differentiating between kinds of tools? Well, it is not clear that we obtain a more refined understanding of technology by dividing physical and social mechanisms into separate categories. If something can be used to achieve a goal, it is a tool and, in so being used, can become a technology. Including the bureaucracies in the infrastructure of science acknowledges what successful entrepreneurs of Big Science have known all along.[11] There is a system out there that can be worked to the advantage of science, just as modern laboratories give scientists advantages over scientists working in primitive conditions. Laboratories, funding agencies, etc., are all tools scientists use to advance their research.

In short, as philosophers leave the tidy world of reconstructing the logic of concepts and enter the real world of science as a social process involved in producing knowledge, they are forced to understand the effects of the complicated interplay between theory, experiment, scientists, bureaucracies, and the material world, among other crucial factors. And if they acknowledge all those factors, then they should be forced to rethink the old assumption that technology is merely applied science and, as such, applied knowledge. This is not to deny the fact that from certain scientific discoveries we have been able to produce devices that make life easier. That is, some technologies *are* the result of applying the knowledge that science produces. But that does not exhaust the characterization of technology, since technologies are also involved in the production of scientific knowledge itself, not merely in the context of experimentation but also in the forms of the institutions of science, among other things.

Section 3. Defining "Technology"

To accommodate this social dimension, I am proposing an account of technology that takes us beyond the standard notion of tool-as-mechanical-mechanism. This is essentially a redefinition of "technology." And if I am

10. See Mumford 1963.

11. For the distinction between Big and Little Science, see Price 1963.

to be engaged in the dangerous process of attempting to define "technology" in a new way, then it seems both appropriate to bite the proverbial bullet and worthwhile to include as many dimensions of technology as possible. One way to accomplish this is to expand our account beyond the more standard view of tool-as-mechanical-mechanism to tool-as-mechanism-in-general. Thus, if a tool is conceived of as a means to facilitate accomplishing some task or other, then it would appear that governments, organizations, and hierarchies should be construed as tools just as hammers and nails are.

For too long technology has been characterized only as mechanical tools, despite the fact that not all the tools humanity has used to secure itself against the elements are mechanical ones. Some of humankind's earliest and most important tools were the social structures it devised or evolved for establishing order and protection. Surely these need to be included in a good account of "technology." But extending this understanding of tools to incorporate these social phenomena must be done carefully. Some see the following danger: that if construed too broadly "technology" may become a concept without content. But what some see as a danger may be a benefit. I will return to this issue later. Nevertheless, to understand our technologies and how they relate to science and society in general, "technology" must be seen in broader terms than have been used previously.

Others, such as Emmanuel Mesthene, have seen this point as well. In his short brief for a commonsense approach to understanding technology, Mesthene characterizes technology as "the organization of knowledge for the achievement of practical purposes" (Mesthene 1970, 25). This characterization permits the latitude I seek, but it still needs some critical attention. For instance, in an epistemology that rejects focusing exclusively on the beliefs of individuals, the phrase "the organization of knowledge" is redundant. However "knowledge" is defined, it eventuates in a structured set of claims, for which organization is a necessary condition; gone are the days when knowledge was conceived of as an unordered set of unique assertions standing apart from all external influence and independent of one another. That being the case, Mesthene's account of technology comes out sounding like "knowledge for the sake of practical purposes." Not only is this a perfectly acceptable account for a pragmatist, but it also provides an initial structure for unpacking a model of technology that both exhibits the insight uncovered here and permits exploration of the nature of the similarities in the varieties of technologies I wish to consider.

But before I continue, I want to take one more look at Mesthene's revised definition. Earlier I argued that there was no such thing as knowledge for its own sake by way of attempting to undermine the idea that science could be characterized as pure while technology was merely applied knowledge. In effect I was arguing that all knowledge is for practical purposes. And if that is the case, then my revision of Mesthene's definition of technology to "knowledge for the sake of practical purposes" again becomes redundant since it was already established that all knowledge is for the sake of practical purposes. And if that is taken into account, the following definition of technology emerges: technology is knowledge. And while it sort of sounds good, it doesn't reveal much. Nevertheless, I hate to give up Mesthene's insight because it seems so promising.

The way out of this dilemma is to draw a distinction. The important part of Mesthene's account is his insistence on the *use* to which the tools we invent are put; *technology is using tools*. That seems to capture some of the crucial factors, but not all, since it is not clear that the use of tools is necessarily goal-directed, and it is the use of tools for specific goals that should be included as the foundation of our definition. Furthermore, there is a certain sense in which our account of technology should refer to the sorts of things humans do, at least for now. Later the definition might be expanded to include the activity of beavers or aliens—but before that happens it is essential to have a good idea of purposeful activity for non-humans. This will be no small task, for the fact of the matter is that a good idea of purposeful acitivity doesn't exist for humans yet. So for the present let us restrict our attention to human technologies.

With these two ingredients in hand, namely, that it is the activity of humans that is the object of concern, and that it is their deliberate and purposeful use of tools, taken in the general sense, that characterizes technology, I propose the following definition: *technology is humanity at work*. But definitions by themselves do not solve conceptual problems such as the ones I have been exploring about science, knowledge, and technology. Good definitions are best elaborated and understood in specific contexts. In this case, I propose in Chapter 2 a *model* of humanity at work to provide the means for unpacking the definition and giving it a context in which to make sense.

As suggested above, it might be objected that this definition is too broad, too inclusive. Under this characterization, everything we do, other than for play or leisure, becomes technology. This is the worry raised earlier, that too broad a definition yields one with no content. If everything, including science, becomes subsumed under the rubric of technology, we have a useless definition, since the purpose of such definitions is to help us make distinctions. This definition makes it impossible to draw any distinctions; hence, it is argued, it must be abandoned.

I disagree that we should abandon the definition for three reasons. First, this definition has the distinct advantage of pointing to the futility of talking about technology *simpliciter*. Or, put another way, maybe any broad conception of technology is, in fact, empty. There is no technology *simpliciter*. Second, as noted earlier, technical definitions cannot function well with an accompanying theory or model. So I urge that before rejecting the definition, we should check out the model.

Finally, this definition allows us to make the distinction we need, which is between the tools and their use. The tools themselves are not the technology; it is the use to which they have been put that marks out a technology, and it is people who do the putting to some use for some purpose. So now, with a definition in hand, it is time to see what kind of a model it suggests and how this will help deal with the philosophical problems technologies pose.

One last word: despite my efforts to develop a new definition of "technology," one that focuses on what people do and how they go about doing it, there will be occasions when it will seem appropriate to talk about "technology," where it is the tools, techniques, and systems of tools and techniques that are at issue. This second use of "technology" should be clear from the context. In using the term to refer to tools, techniques, etc., I will be trying only to capture the more popular conception of "technology," not to endorse it.

CHAPTER TWO

Modeling Humanity at Work

COMMON TO TOOLS, INSTITUTIONS, and decision-making procedures alike is the simple process of transforming some input into an output. In making a decision, knowledge is part of the input and achieving a practical outcome is the output. Following this line of thought, in the effort to provide a context for the definition "technology is humanity at work," leads to a model of technology as an input/output transformation process.[1] There are also input/output transformation processes whose function is to develop other input/output transformation processes. Let us distinguish between these as first-order and second-order transformations. Decisions are first-order transformations. The result of a first-order transformation either may be another first-order transformation, i.e., a decision to make another decision, or it may lead to a second-order transformation, i.e., a decision to create a tool of some sort.

A second-order transformation involves a constructed device. An oil refinery performs a second-order transformation; so does a legal system or a geometry or a telescope. They are the results of first-order transformations in which decisions were made using available knowledge, etc., to build a refinery, to adjudicate conflict, to construct a system to measure special relations, or to see faraway objects. Thus, decision-making procedures are first-order transformation processes or first-order transformers. In the case of the construction of a refinery we have a nicely complicated example, because the decision to build the refinery actually amounts to authorizing another series of first-order transformers, which are the processes to be used for planning, designing, and constructing the project. The building of the refinery involves decisions as well as the manipulation of materials. The completed refinery is itself a second-order transformer

1. The basis for this model, let us call it MT, is Glendon Schubert's analysis of the structure of decision-making in the U.S. Supreme Court (Schubert 1965).

since it transforms raw materials by mechanical means. So, using the basic notion of an input/output process, we can still distinguish among mechanical, social, and decision-making processes, thereby allowing such institutionalized decision-making processes as bureaucracies and funding agencies to be characterized as technologies.

But to characterize technology as humanity at work, meaning by that the set of first- and second-order input/output transformations, doesn't mean that it can't be analyzed further. To complete the model we need more than the notion of levels of input/output transformations. There is a crucial third ingredient we must include if our model is going to reflect the most important component of humanity at work, where that work entails some sense of making progress. The final ingredient is assessment feedback. Let us consider each of the three components again.

(1) A first-order transformation process is a set of deliberations wherein, using an already established knowledge base or starting from a given state of development, we confront a decision-forcing situation generated by what is perceived to be a problem or a set of problems. Some of the solutions to these problems may require the creation of new machines. Other problems may require for solution the implementation or elaboration of a specific kind of social machinery (a committee, or a new organization, a company, a bureaucracy, or a legal system). Whatever the nature of the particular solutions to these kinds of problems, the setting proposed for those deliberations raises a number of other issues. Among these are the nature of the deliberative process, the structure of practical reason, the nature of rationality, and the definition of "knowledge." We will return to some of these later.

(2) The second part of the model consists of secondary processes that exhaust what is normally called technology in ordinary usage, i.e., the machines. But it is also a broader category, including, as it does, social structures. Thus, an oil refinery can be discussed in terms of the transformation of crude oil into petroleum, or a legal system in terms of the transformation of conflict into resolution, or science in terms of the transformation of knowledge, theory, instruments, and data into more knowledge. It is important to note here, however, that whatever the end product of the process, it is not an end in itself. These products have specific purposes and, therefore, further uses.

Despite our insistence that there are no ends in themselves taken simply, there remains, nonetheless, a venerable philosophical distinction between intrinsic and extrinsic ends. Intrinsic ends are supposed to be ends in themselves, as opposed to extrinsic ends, which are items or results concerned with furthering other ends. But like other distinctions under attack here, this one won't serve us well, since it too proves too limiting. The limitations become apparent when we realize that social machinery offers us a kind of product different from that offered by mechanical machinery. If we construe the end product of the legal system as justice, justice can be seen both as an end in itself and as furthering social productivity by providing a reliable means for mediating conflict. But unless you have already agreed on the basis of a set of values that have been adopted prior to any serious contact with the world (e.g., that justice is somehow an end in itself), the argument in favor of a legal system producing justice as its product can be sustained only in terms of its social worth.[2] Likewise, if science is a social process whose product is knowledge, given our earlier discussion we can't argue for knowledge as an intrinsic end.

(3) The third and final component of our model of technology is assessment feedback. Technology assessment is a special kind of decision making in which the effects of implementing decisions of the kind in (1) above are illuminated by means of a feedback mechanism that makes it possible to upgrade the knowledge base for further decision making. The scientific process, as one aspect of humanity at work, may be its best general example. In science the constant reassessment of theoretical assumptions in the light of new results, both empirical and theoretical, is essential to the continuing development and refinement of scientific theories.

But more to the point, the most important aspect of contemporary technologies is the extent to which an assessment feedback mechanism is formally incorporated into the decision-making procedures surrounding the development and implementation of plans for new ventures. In some cases this is mandated by government. But even more interesting is the situation today in which the importance of assessments and feedback loops is being insisted on because of the magnitude of some technological ventures and their potential consequences. It even appears that assessment has become an important value governing humanity's work. A full-scale discussion of this phenomenon (not to be undertaken here), would reveal the means by which changing goals and values affect the development and

2. This is a contentious claim, for many theorists argue differently. I cannot pursue the matter here, for our interests lie elsewhere, but I would note only that this account of justice is at least consistent with a thoroughgoing pragmatism.

implementation of new and innovative techniques for transforming raw materials into suitable results in both the physical and social domains.

At this point it should be apparent that MT is intended only to schematize the complexity and pervasiveness of technology rather than to be a definite description of its structure. On any detailed analysis it will become clear that wherever a rational decision is involved, so too are a variety of other considerations, among them the assessment process, the nature of the second-order transformer, i.e., mechanical or social, the social circumstances, the goals of the individuals as well as the goals of the institutions, and so one cascades down a virtually unclosed spiral.

The success of this model, then, should not be judged in terms of how it simplifies our view; it doesn't, nor was it intended to. The world is a very complex place, and we do ourselves serious disservice by thinking that we can always get around in it better by simplifying things. The merits of the proposed account are to be found in the manner in which it allows the complexity of any actual situation to be exhibited, while still providing the means by which to isolate and analyze the relevant components and their interactions in the light of constantly changing circumstances.

Section 1. Practical Reason and Rationality

While briefly describing the first component of the social process model of technology, i.e., first-order transformations, I enumerated four items that need amplification: (1) the structure of the deliberative process, (2) the structure of practical reason, (3) the nature of rationality, and (4) the definition of "knowledge." Concerning the structure of the deliberative process, a great deal of work has been done by a number of economists and general theorists of decision making, to which little can be added here. I have already initiated a discussion of the problem of knowledge and I propose to leave that aside temporarily, preferring to return to it at the end of this chapter and again in Chapter 3.

Let us now turn to the structure of practical reason and the nature of rationality. My view of practical reason places discussions of rationality in a very different light from more standard accounts. In partial defense of what some will see as an unorthodox view, I will show that it should also militate somewhat against absolutist judgments on the merits or demerits of this or that technology.

It is difficult to separate out the intertwined strands of rationality and practical reason. Talk of one almost always brings in the other; so too here. By way of working toward an adequate view of practical reason, we might be able to see where the proper distinctions should be drawn by considering first what I propose is a flawed theory of rationality that gets the two concepts mixed up.

This account characterizes a rational person as *homo economicus*—i.e., an economic actor programmed (socially and/or genetically) by the principle that one should maximize one's gains and minimize one's losses (or some variation on this theme).[3] The basic idea here is that people are motivated to act primarily through self-interest and that their self-interest can be determined by finding out what states of affairs they would prefer to see come to pass. An individual chooses between alternative outcomes on the basis of determining which outcome is preferred. In order to know what is "preferred," the individual has to rank-order his or her preferences. Not only is it difficult to determine what one's preferences are, but determining their rank order is even more difficult because the order changes, depending on circumstances and on the state of the actor's knowledge base (Minas 1973). The characterization of these preferred states of affairs is varied; sometimes they are referred to as "values," sometimes "revealed preferences," sometimes "utilities." But whatever they are called, on the *homo economicus* model the rational person always acts to maximize gains and minimize losses, or minimally, to maximize perceived gains.

My complaints with this approach to rationality are two: (1) on an individual basis, that is, for the purposes of explaining the behavior of a single individual, the *homo economicus* model is too successful—like Freudian psychology, it can be used to explain everything; (2) on the other hand, at the level of group decision making, it can't readily explain very much. Group interests are hard to identify without generating self-satisfying criteria. Despite these problems, the *homo economicus* model is usually defended as coming fairly close to representing the way things actually are.

The difficulty with the individual case is that a plausible *homo economicus* explanation of why an individual acted as he or she did can always be constructed *after the fact*. That is, after the fact we can always say that the individuals in question did what they did because they thought they were going to maximize these gains, etc. This way it doesn't matter if they selected the choice that actually gained them their optimal scenario; all that matters is that their behavior can be construed as acting as if they believed their choice would do so.

The situation, however, is not so easy when one is *predicting* individual behavior, since predicting actual individual utilities ahead of time is hopeless; preference rankings and the reasons for changes in such orderings will not necessarily be accounted for on the basis of rational principles. That is, rational considerations are often not the determining factors

3. See Bonner 1995, for a history of the connections between utilitarianism and economics. Also see Hausman and McPherson 1996, for an analysis of the connections between moral theory and economics.

in why person x prefers a to b. Religious fervor, for example, might lead a person to place himself deliberately and with complete understanding in a life-threatening situation. *After* the fact, self-sacrifice by religious fanatics can be argued to be as rational as unloading stock on Friday before the market closes after a week of rising prices on the stock market. All we have to do is to assume that the individual has rearranged his or her preferences so that self-preservation has been replaced at the top of the list by the desire to achieve, for example, religious fulfillment. In this way all decisions become rational in retrospect. This, however, deprives the concept of rationality of its value. Rational actions are supposed to be the ones that are preferred because it has been determined ahead of time that these actions will yield the desired effects. If any action turns out to be characterizable after the fact as rational, then all actions are in principle equally preferable, which is surely not the case.

To some extent these problems can be minimized. For example, to shore up the predictive ability of the individual model requires that we be able to predict better what people will do. One way to increase our ability to make accurate predictions of this kind is to increase the kind of information we factor into our predictions. Thus, for example, if we could devise a method for correctly categorizing individuals according to membership in social groups, we would thereby facilitate isolating some of the relevant governing social norms that sometimes function as overriding principles for individual choice.

The assumption here is that whatever goals and objectives an individual may have, those goals are generally obtained within a social context, thereby requiring that actions designed to achieve those goals accord with socially accepted norms. This is where we can register the point made earlier about knowledge as the product of group endorsement. Social norms are not merely utilities to be weighted and figured into some formula. Rather, *social norms function as setting the context* in which further deliberations can occur. This is true even for the extreme case such as the rugged individualist. For example, rugged individualists can be identified as belonging to a group whose behavior can be characterized in terms of certain general principles of action, however antisocial or nonsocial that may be. Under these circumstances, if one is to continue to be classified as a member of the group, only certain options for action would be considered appropriate. So, to the extent individuals continue to use those principles, we increase our ability to predict their behavior and we learn more subsequently about the behavior of the rugged individualist in general. Consideration of group membership thereby narrows the set of possible options in any given decision.

This general account can be made more precise by distinguishing between the *general reasoning pattern* which sets the *context* for reasoning and, on the other hand, the reasoning which generates or precipitates a particular action.[4] The context is set with the selection of an appropriate policy to guide one's actions. The general form of such a policy statement is:

When in circumstance *C* do *X*.

Having determined that one indeed is in circumstance *C*, then one is entitled to believe that he or she is warranted in deciding to do *X*.

Deciding whether or not one is in the appropriate circumstance is not usually deemed to be a problem. But where technology is concerned, it is the *basic problem*, because it is the first one over which difficulty arises. It often *appears* that disagreements over technological innovation are arguments over the consequences of taking certain actions, but that is often only because people disagree over what they understand the circumstances actually to be. Once having determined what the circumstances are, then the consequences of acting in certain ways can be clear, or at least clearer. It is at that point that the real trouble starts. For even in those cases where we know fairly well what will follow, we can rarely agree on what we ought to do, because decisions about what we ought to do are based also on values and goals, not just the facts of the situation. Thus, assuming that we all agree that *X* is the amount of residual radiation we are exposed to in standard circumstances (however defined), nothing follows. The conclusion concerning what to do, granted we are in *X*, will be a function of whether or not *X* meets an acceptable level of risk, and that decision will be a function of a host of other factors, some factual and some not.

Section 2. Rationality

In order to select the best thing to do out of a range of options, we need a means for determining what constitutes the "best" option, i.e., what, on normal accounts, is the rational thing to do. The *homo economicus* model provides a mechanism by which we can find out what individuals may think is the best option for them. But our problem is not resolved by having a method for determining individual choices. What is best for one person is not obviously best for all or even for a large segment of the populace. The key issue in front of us is the problem of coming up with a method for resolving the conflicts that arise once individual choices have been made, for when it comes to making decisions about technology, the major problem is resolving disagreements over the appropriateness of some new machine or system, and this resolution takes place at the social

4. This is the line of attack Wilfrid Sellars takes in his "Induction as Vindication" (1963b).

or political level, not at the individual level, which is not to say that individuals don't make decisions about which technologies they will use for their private purposes. My concern here is with how we arrive at social policies regarding technologies in the public sphere.

Thus the question of the rationality of the choice of what to do is no longer directed to the individual so much as to the community of actors whose role is to decide the general policy that in circumstance *C* it is best to do *X*. These sorts of decisions are rarely straightforward. If anything, the factors that bear on group decisions are seldom matters of pure reason or logic alone. Thus, to draw an analogy between the social community and a board of directors for a major corporation, given that the board of directors agrees that its job is to maximize profits, little follows as to how this is best accomplished. Indeed, in the minds of the members of the board, even what constitutes maximizing profits may vary. Some members may concentrate on profits calculated strictly on a yearly basis, while others may insist on taking a long-term view and look for trends over a five-year period. In the ongoing debate over the possible adverse effects of the policy of rewarding top executives with bonuses based on yearly profits, it has been suggested that this policy encourages short-term profit-taking at the expense of the long-term health of the company. In other words, the bottom line of the Annual Report may not be the bottom line. How a board of directors is supposed to resolve this issue on rational, i.e., *homo economicus*, grounds is not clear, since the difference between their preferred choices is a direct function of their individual values. Thus you would have to convince someone to change his or her values in order to get him or her to agree with you, and this is rarely, if ever, achieved by appealing to rational principles. Even if you appealed to a person's rational self-interest, the use of such a principle is ultimately self-defeating, since it requires detailed knowledge of each person's preferences and their rankings, and those will change, especially when the agent realizes that it is his interests that he is being urged to consider. Every once in a while you may get a person to adjust his or her values so as to have them come to a conclusion more to your liking by suggesting that if they do so they will actually come out better in the long run, but this method does not guarantee success, and, furthermore, it simply extends the problem since it concentrates on the individual's needs rather than those of the community.

Maybe here too the problem stems from employing an outdated notion. Anyone who has ever engaged in group decision making knows that the stark ontological primacy of compromise is the furnace in which all general policies for action are cast. Why then should we assume that such results themselves will be the result of rational procedures? The stumbling block to overcoming this roadblock lies in the nature of the burden we place on the notion of rationality. On all the standard accounts of rationality, a rational choice is supposed at least to increase significantly the probability that an agent will make the right choice. The right choice is supposed to lead to the best possible results. A rational decision is one that puts the actor on the road to success—or so it is supposed. If these results are to come to pass, then it must be the case that there is something behind our rational decisions that guarantees this result. In other words, appealing to rationality is an appeal to some sort of guarantee. This is the sort of guarantee that says something like, "if you proceed in this way, you have the best shot at arriving at the best result." In other words, appeals to rationality substitute for a *guarantee of success*. And where, one asks, would such a guarantee come from? Traditionally there are two sources: logic and knowledge.

Unfortunately, logic can guarantee only the rigor of an argument. It cannot establish the truth of the premises in that argument. So an appeal to logic is only as good as the truth of the initial claims. The most "logical" argument can still lead to incorrect claims about the world if the premises are false to begin with.

The case is not much better for knowledge. Both the rationalists and the empiricists (with the exception of David Hume) attempted to develop a definition of knowledge that equated knowledge with certainty. They sought a secure foundation upon which all knowledge claims could be developed, with the foundation providing the guarantee of certainty. David Hume took a different approach—he showed that on the new approach that equated certainty with having a clear justification for having specific ideas in your mind, no certainty was possible, thereby undermining that entire program.

Ever since Hume, modern epistemologists have been trying to develop an account of knowledge that recognizes the fact that each claim we accept today as knowledge may be replaced tomorrow by a different claim. It matters not that they have yet to succeed; what is important is that on virtually all contemporary epistemologies it is now agreed that what we call knowledge today cannot provide the kind of grounding that would provide the guarantee for a theory of rationality. *The problem here lies in the idea that being rational is supposed to enhance your chance of success,* and if knowledge is uncertain, then the payoff for being rational is diminished.

The way to see how rationality is linked to success is to consider a paradigmatic case of being *irrational*. We say of a person that he or she is irrational when he or she deliberately chooses to do something that he or she knows will not give him or her the results he or she really wants. If he doesn't know what he is doing, then it is a case of acting on misinformation or out of ignorance. But if he does know what the outcome will be, and if that outcome is contrary to what he really wants to take place, then we say he is irrational. The irrational person is faulted because, for the most part, he knows, in the sense of having had the relevant experience, what works and what doesn't. Consider the case in which a person attempts something and fails and then attempts exactly the same thing again, making no changes. This is acceptable for the first retrial. But if the same action is attempted a third time without any attempt or concern to change assumptions, goals, methods, etc., then we have a problem. More to the point, this is clearly an irrational move. But unlike traditional accounts that would characterize that person as irrational because he failed, I wish to argue that he is irrational not because the action failed to achieve the desired result, but because *the person failed to learn from the previous experience*.

Accepting this approach to understanding irrational actions provides us with the way out of our problem with defining "rational." The ordinary sense of rationality offers promises of success, with the ideal account guaranteeing success. But success is impossible to guarantee. On the other hand, if we cannot guarantee success, we can at least avoid making the same mistakes twice. That is our clue. Here then is the solution to our problem:

> The Commonsense Principle of Rationality (CPR): *Learn from Experience.*[5]

A nice feature of CPR is that it separates being rational from being successful. If you learn from experience and avoid repeating the same mistake, it doesn't follow that you are guaranteed success. Success means achieving a previously determined objective, and that can be accomplished by systematic means, luck, or a combination of both. Mechanized procedures can be successful. It doesn't follow that they are rational. Machines and procedures aren't rational; people are. (And, gentle reader, if you are tempted to use that aphorism as a justification for claiming that computers can't think, I still defer to the Turing Test.)[6]

In addition to providing us with an account of rationality that is neither trivial nor self-defeating, CPR also has a methodological dimension. The basic idea requires recognizing that the acquisition of knowledge is not the simple linear process of discovering truths one at a time and adding them to the open-ended encyclopedic bag of truth. Rather, the acquisition of knowledge is a continuing, ongoing, dynamic affair, the analysis of which requires the use of a feedback loop. What we know at any given point in time is the result of previous stages of acquisition and

---

5. I first introduced this account of rationality in Pitt 1991.

6. For recent work regarding Turing, see Millican and Clark 1996; and Clark and Millican 1996.

refinement. This includes values and goals as well as information. The more information we acquire, the more we are forced to sort it out against the background of what we already believe we know, i.e., what information we have already acquired. But we do not simply proceed by acquiring new information and storing it on equal footing with what is already there. As we learn new things, we are forced to consider the effect of that information on what we already believe.[7] In many cases this is not immediately apparent and comes to the front only when a series of experiences combine to yield a new understanding or perspective at odds with our current epistemic state. Thus our reconsideration of what we know may be delayed because the consequences of new information for what we think we know may not be immediately obvious. In so doing we are constantly rearranging the content of our beliefs and, in that light, we are also forced to reconsider the merits of our goals and the appropriateness and structure of our value system. When we stop doing this, we stop learning. To avoid repeating mistakes, i.e., to be rational, requires that we learn from those mistakes, that we update our information, eliminating incorrect data and assumptions that led to failure, and then try again, reevaluating those results, and continue in that fashion.

CPR, as it turns out, is also nicely compatible with our input/output assessment model (MT) of technology. MT provides for assessment simply because it is a feature of technological growth. As a normative principle, CPR, on the other hand, *requires* assessment, upgrading, and reassessment. With CPR we assess technology, i.e., humanity at work, to determine whether or not progress is being made. And that progress is constantly measured against what we have come to know about the feasibility of our goals and values.

Section 3. Conclusions

So far we have managed a working definition of "technology," a sketch of the technological process, and an account of rationality. These results have come at the cost of reconsidering and even abandoning a number of longstanding assumptions about knowledge, rationality, science, and even technology. Technology is now to be conceived as a complicated process of humanity at work in which knowledge gained by prior action is reconsidered in the light of new knowledge and new actions attempted by way of focusing on achieving specific goals. The process involves assessment and feedback, action, and analysis. It does not promise success, but it offers a methodological program for avoiding the same errors and an

7. We are "forced" to consider the implications usually only when the new information conflicts in some way with what we currently know.

injunction that forces us to reconsider what we know, how we are using that knowledge, and where we are going.

But the question of the relation between science and technology remains unresolved. I have characterized science as a knowledge producing process and spoken of technology as a process that uses knowledge. Hence, despite earlier disclaimers, it is beginning to look as if we have not yet escaped from the view that technology is applied science. I have suggested that in the process of doing science, the institutions of science are used as tools. This does turn things around somewhat. But that is still not enough to get the relation right. The final characterization of the relation between science and technology has to reflect the fact that progress in modern science requires a technological infrastructure. This means that the complicated relationship between science and technology involves more than science using some aspects of the institutions supporting it as tools. These institutions, among other components, are essential for science as we know it. That means the case to be made will have to show that technology is epistemologically *prior* to science, which makes perfectly good sense if you just think about it. Knowledge is not possible without some prior actions, and since technology is humanity at work, it naturally follows that technology is epistemically prior to science. And while it is easy to say, it is more difficult to show.

CHAPTER THREE

Locating the Philosophy in Technology

WE HAVE BEEN SEARCHING for the best way to think about technology. And we have made some headway. But we have not yet resolved the relationship between science and technology. The kind of relationship we seek to expose is *conceptual*. That is, the understanding we seek is not to be found in the examination of the history of technology, nor in the history of science, since the proper construction of such histories requires prior answers to the sorts of preliminary conceptual questions we are asking here. To see this, just reflect on the fact that you cannot write a history of science unless you already know what counts as science. Likewise, we cannot explore the science/technology connection until we understand the conceptual components of the relationship and how they bear on one another.

To resolve the conceptual relations between science and technology, I propose the following approach. Much of what is commonly thought about the hard-core nature of science comes from philosophical work around a set of issues primarily identified by such philosophical thinkers about science as the logical positivists. In searching for an understanding of the relation between science and technology, I will start by enumerating some of those same topics and then proceed to see if there can exist reasonable subsequently *counterpart technological concepts*. Basically, my aim is to show that the tendency to link science and technology should be resisted, since it does a disservice to technology. This is because the counterpart technological concepts differ in substantive ways from the fundamental concepts that are constitutive of our notion of science.

Section 1. Counterpart Questions

The use of the expression "science and technology" is so pervasive that we might be tempted to eliminate the spaces between the words and use "scienceandtechnology" in its place, or perhaps just plain "S/T."[1] Despite the constant conjunction of these two terms in current common discourse, and despite the obviously close connections between science and technology (I will leave aside the chicken and egg question, although, if I am right, technology came first) and the obvious association of one with the other in general discussion, when it comes to philosophical discourse there is a startling lack of symmetry with respect to the kinds of questions that have been asked about science and the kinds of questions that have been asked about technology.

In the twentieth century, philosophical questions about science have been dominated by such issues as:

- the nature of scientific knowledge
- the nature of scientific explanation
- the function and structure of scientific laws
- the form and character of justification in science
- the structure of scientific theories
- the nature of scientific change
- the nature of scientific evidence
- the debate between scientific realism and antirealism

What is strange is that it is rare indeed to find *counterpart* questions and issues to those questions about science listed here in discussions about technology. In part to remedy this situation, but also by way of finding a more satisfactory entrée into the philosophical questions concerning technology, I suggest we investigate a set of counterpart questions about technology corresponding to the fundamentally epistemological questions that have been raised about science over the past century and a half. These questions include:

1. The sociologists of science and technology have introduced the term "techno-science" to capture this insight. In the light of the comments that follow, this term will be seen to shed no light on our problem.

- What is technological knowledge?
- What counts as a technological (technical?) explanation?
- What is (are?) the structure(s?) of technological theories?
- What is the nature of technological change?
- What are the function and structure of technological laws?
- What is the debate over technological determinism?

To focus on these issues rather than on social criticism is to adopt a larger agenda for studies in the philosophy of technology than that of the social critics, one in which social criticism is removed from center stage. It is not that social criticism has no place in the philosophical discussions of technology. Rather, as noted earlier, it can come only after we have a deeper understanding of the epistemological dimension of technology, if the subsequent discussion of the social dimension is to merit our attention. If we *begin* by worrying about the good or bad consequences of technology, questions concerning the nature of technological knowledge or the structure of technological theories that ought to form the basis for deciding these issues may never have a chance to come up, and so those decisions must remain *ad hoc* and unjustified.

To highlight the kind of concern here, consider a simple example. It is not uncommon in public hearings held to determine the advisability of a project involving a large-scale technology, such as a high-voltage power line, to have advocates and opponents testify. Each group will bring its own group of scientific experts or appeal to its favorite studies to bolster its case. In some circumstances we have scientists contradicting each other. Clearly, before a reasoned assessment of the merits of one case or another can be made, we need to know how to determine the accuracy of the scientific claims. For example, before assertions regarding the impact of putting a high-tension line across a national forest can be endorsed, the scientific basis for the claims must be evaluated, and this requires, among other things, a good theory of confirmation by which to evaluate the kind of support the alleged evidence gives to the conclusions. Once we are sure of the facts of the case, then, and only then, can we proceed to debate the advisability of a proposed course of action. Thus policy decisions require prior assessment of the knowledge claims, which require good theories of what knowledge is and how to assess it.

Furthermore, unless the epistemological issues receive prior analysis, much of the social criticism will miss the mark. For example, in the debate over the safety of nuclear energy plants, how often have the various conflicting assessments of the technologies involved differing assumptions about the nature of technological knowledge? On reflection, we might make a case for understanding much of the debate in terms of differing epistemological assumptions: those opposed to nuclear plants seem to be assuming that the kinds of knowledge generated by assessment procedures should meet the standards of an idealized form of scientific knowledge, while those who argue that the technologies of nuclear energy provide adequate protection seem to be employing a different set of epistemological standards, i.e., standards derived from the practice of engineering. Since one of our objectives is to integrate philosophical discussions of technology into the larger philosophical dialogue, we should not bypass or ignore important dimensions of the topic under discussion. If, by taking the social/critical turn, we block the possibility of raising the epistemological and metaphysical questions in a systematic and relevant manner, we can and do, in effect, undercut the philosophical effort.

Section 2. Technological Knowledge: Setting the Stage

Whatever science and technology are, they must be understood as historical phenomena that must be seen in the specific *sociohistorical contexts* that give them their distinctive characteristics. Since social contexts change over time, we should therefore assume that science and technology, seen as historical phenomena in changing social contexts, will themselves change. It is not the purpose of this book to provide a full or even a partial history of science and technology.[2] My account of technological knowledge will be restricted, by and large, to technology in its most modern guise, taken roughly as the period since World War II. Hence much of what we have to say about technology may not apply to earlier historical periods.

Just as science and technology are undoubtedly sociohistorical phenomena, there can be no doubt that science following World War II is significantly different from science prior to World War II. The difference lies primarily in its organizational structure. World War II, marking the difference between Big Science and Little Science,[3] designates the change from science popularly seen as the domain of individuals blindly pursuing "truth

2. But in an effort to flesh out the historical bones of this claim, there is my work in progress: *Seeing Near and Far: A Comparative History of the Development of the Telescope and Microscope*.

3. Price 1963.

for its own sake" to contemporary science, the science of big government funding and university-entrenched research programs. This establishment institutional-based science is more goal-directed, more influenced by funding opportunities made possible by governmental and social factors that, until recently, were normally considered by philosophers but not by historians and sociologists as external to the scientific process itself.[4]

The philosophical analysis of scientific knowledge has proceeded, by and large, as if the change from individualist science to institutional-based science had no effect on the nature of scientific knowledge. These assumptions are being challenged by recent work in the sociology of science. The results of the debate are not yet clear, but, following the discussion in Chapter 2, one thing is certain: the primitive epistemological assumption that knowledge is what individuals create on their own must be abandoned. Both the determination of the criteria for knowledge and the endorsement of specific claims as knowledge are community activities. But even in the age of Big Science, where large institutions play such an important role, science remains at rock bottom an activity constituted of the work of individuals. Individuals produce candidates for community acceptance, but individuals alone do not produce scientific knowledge. What counts as knowledge is ultimately the product of a long process deriving from scientific research. From the research process scientists propose results for public scrutiny and for publication. Other scientists determine if those results fit into accepted standards and theories, and evaluate their demonstrated usefulness. Only when the community of evaluating scientists endorses the candidates that individuals have proposed based on their research do we have a product called "scientific knowledge." Even so, the process continues and what was accepted may eventually be rejected in the light of new findings. This, however, does not mean that because science involves a social process, it is "social all the way down."[5] Science is about the world. Its objective is to explain why things in the world behave as they do. That the process by which we come up with explanations is a social one does not entail that there is no fact of the matter, no world, that plays a crucial part.

Understanding the complicated process whereby scientific knowledge is created helps explain how scientific knowledge can constantly be changing. All scientific claims are up for grabs precisely because the needs,

4. Merton 1970; Merton 1973; Westfall 1958.

5. Bloor 1991; Collins 1985; Latour 1987.

methods, and criteria of the community are changing in the light of factors both internal and external to the scientific process itself.[6]

What about technological knowledge? Can we discuss it in a fashion parallel to what we have just been saying about scientific knowledge? Can we distinguish Little Technology from Big Technology, and can we locate the transformation at World War II? Was there a similar transformation in the technological domain from individualistic technology to institution-based technology? Is technological knowledge a function of community endorsement according to standards and needs that change over time? Let us examine some of these issues, proceeding in reverse order.

If we are to carry out the counterpart concept strategy, we first need to identify an appropriate technological counterpart to the community of scientists who are responsible for the production of scientific knowledge. In the popular mind, technology has long been associated with things, artifacts. But artifacts are themselves the final result of a process. Thus, if we are to locate a parallel to the scientific community, we should isolate a process that leads to the production of the artifacts and identify the individuals and their roles in that process. Having identified the process, we can then ask if the nature of the process changed around World War II, etc.

Disassociating technology from artifacts and concentrating instead on the process that produces the artifacts also retains the spirit of the characterization of technology in Chapter 2. Since I am looking at the process that produces the artifacts, rather than the artifacts themselves, and since the production of artifacts is the result of humanity at work, I would suggest this is a coherent approach. But it won't do without an account of "work."

There are many different ways to approach the concept of work. I propose to characterize work as *the deliberate design and manufacture of the means to manipulate the environment to meet humanity's changing needs*

6. This is a *fallibalist* view of scientific knowledge. It has been criticized because it suggests a general, relativistic position. This need not be the case. One attempt to block the relativist slide is to suggest that while scientific knowledge may change, it has a final touchstone in reality that cannot be avoided, i.e., *realism*. I do not intend to enter into the realism/antirealism debate here. But it seems that one need not appeal to realism to stop the total relativism of what has been called the Strong Programme in the sociology of science. One can agree that scientific knowledge changes over time with respect to what we know and what we want to achieve, without accepting the conclusion that knowledge is nothing more than the results of social agreement. This history of epistemological failure, the restraints of fitting together the complex totality of what we know at any given point, and the bare facts of the world itself are sufficient to stop any such slide. Anyone who argues otherwise is playing games and is obviously more interested in "making points" than in understanding the phenomena under discussion. Academic arguments over points of logic can be fun, but they should not distract from the business at hand.

and goals.[7] The key here lies with the notion of design. The appropriate technological counterpart to a scientist must, like the scientist, be both the creator/initiator and repository of knowledge. And, just as the research scientist must be equipped with the knowledge of how to manipulate nature in a way that allows its secrets to be revealed, the technologist must know how various mechanisms work and how to combine them to produce new mechanisms. This is a process of design. So that technological counterpart to the scientist must be more than a skilled craftsman who knows, for example, how to make a waterwheel. The skill in question is more akin to using the knowledge of how a waterwheel works to design a mill. In short, the technological counterpart to the contemporary scientist is today's engineer.

The engineering community is composed of those groups of individuals responsible for the design and manufacture of the artifacts we use to transform our environment, natural, social, and domestic, to suit our changing needs. In this respect it is no accident that, when we talk about making changes in social arrangements, we talk about social engineering.[8] But we will not go into that topic. For our purposes, for now, we will stick to the traditional engineer. This has certain advantages. In particular, we can identify certain social institutions to help us identify who is an engineer, in much the same way we identify scientists. There are, for example, schools of engineering, engineering professional societies, journals, academies, etc. Furthermore, engineering, like science, is not monolithic. There are mechanical engineers, civil engineers, aeronautical engineers, etc. And so, in just the same sense as there is no science *simpliciter*, there can be no engineering *simpliciter*. Having identified the appropriate counterpart group of individuals, we can now attempt to answer the questions we raised earlier about the possible parallels between modern Big Science and modern technology.

On the surface, at least, the answers to these questions are clearly negative. For example, Big Technology—large-scale projects of engineering involving large numbers of people, huge resources, careful design and execution—dates as far back, at least, as the pyramids of Egypt. In that sense, World War II no more marks the beginnings of modern engineering as we know it today than the development of the Salk vaccine marks the beginnings of modern medicine. One might even be tempted to argue that the engineering achievements of World War II, such as the atomic bomb, would not have been possible if engineering had not already been a systematic and developed field of knowledge.

---

7. Under this account, many things people do probably will not count as work. But without going into a detailed economic analysis of labor, I will admit to this and assert that under this analysis, "work" takes on a technical meaning in the context of this theory of technology.

8. Consider the work of B.F. Skinner, especially Skinner 1971; Skinner 1976.

With respect to the second question, did World War II mark a transitional point in the organization of engineering activities? Again the answer seems to be "no." The Ecole Polytechnique in France dates from the 1790s, and the development of American colleges of engineering takes place largely in the last quarter of the nineteenth century and the beginnings of the twentieth century (associated mainly with the Morrill Act and the establishment of that original American phenomenon, the land grant college).[9] The large professional engineering societies were all in place by 1925. Large-scale engineering projects such as the American transcontinental railroad system, the Panama Canal, the Suez Canal, the Hoover Dam, all predate World War II, as does the Industrial Revolution.

The existence of large-scale projects means that while there may have been engineers who practiced their trade in an individualistic mode, there certainly were others who operated in institutional settings. So once again, World War II marks no clear boundary for engineering.

The final point we need to consider concerns the nature of engineering knowledge. Did it change in any significant fashion around World War II, as it appears scientific knowledge did? That, of course, depends on the meaning of "engineering knowledge." And it is here that I can begin the serious effort to integrate questions concerning technology into the larger philosophical story, for now I can begin the process of exploring counterpart concepts to those discussed in the philosophy of science.

The nature, structure, and justification of scientific knowledge has been a topic of central importance for most of the twentieth century. While it is still not clear that there is complete consensus on the criteria for scientific knowledge, several key features have emerged from the discussion. Once those characteristics have been identified, I will examine the possibility of counterpart concepts in engineering knowledge, and then attempt to extend the results to "technological knowledge" in general.

Section 3. Scientific Knowledge

To begin this part of the discussion, let us look at the traditional account of scientific knowledge deriving from the aspirations of the creators of the "new science" of the Scientific Revolution. From that tradition, there are several treasured characteristics of scientific knowledge which recent discussions have forced us to abandon or significantly modify.

Traditionally, scientific knowledge was described as "universal," "true," and "certain." As the special features of the different sciences, most notably the social sciences, become more pronounced, the universality claim has had

9. For a contentious account of the rationale for the creation of the land grant system, see Collier 1998.

to be modified and carefully bracketed. In the social sciences the development of social relativism made this inevitable. Likewise for "true" and "certain." But in these cases the problems are not due to specific aspects of the individual sciences. Rather, they are the result of the difficulty, on the one hand, of demonstrating the truth of scientific claims in a non-question-begging manner, and, on the other hand, of recognizing the fundamentally underdetermined nature of the relation between any scientific claim and its evidence.

Despite these problems, and given some emendations formulated in the light of criticism, the newly reconstituted traditional account offers some features that remain viable. For example, it characterizes scientific knowledge as produced by researchers exploring the domain of a *theory* who aim to provide an account of the relations among the objects and processes of that domain, which account provides the basis for an explanation of phenomena generally observed or detected in another domain. If I were tempted to isolate one crucial characteristic of scientific knowledge, then, it would be this: scientific claims derive their meaning from the theories with which they are associated, hence, *scientific knowledge is theory-bound.*[10]

This situation, the theory-bound nature of scientific knowledge, presents additional problems beyond those noted above for some traditional assumptions about scientific knowledge, in particular the view that scientific knowledge, if true, is true for all time. If scientific knowledge is theory-bound, and if, as we know from the history of science, theories change, scientific knowledge changes; hence, what is accepted as scientific knowledge is not true for all time, at least not all of it, yet.[11] But this should not be a startling claim. The development of human knowledge is a process of continuous exploration in which we reevaluate what we know in the course of new findings, and we jettison what we find no longer consistent with the latest information. Rather than clashing with the history of science, this commonsense view only causes trouble for philosophical theories of the growth of knowledge that ignore that history.

---

10. This is not the place to explore the intriguing question of the relation between theory and the technological infrastructure of science, but it should be noted that there is a complex interaction between the theories scientists employ and that infrastructure. For example, sometimes the examination of the kinds of objects that populate a given domain may be made possible by new instruments, e.g., Galileo's telescope revealing the existence of the moons of Jupiter. Likewise, certain theories may require increasing sophistication in their supporting instrumentation. Referring again to the history of astronomy, once viewing the heavens through the telescope was possible, questions concerning the size of the universe forced modifying the telescope by incorporating a micrometer for the purposes of making such measurements, which required developing a theory of measurement and distance, etc. (See Pitt 1994).

11. I am referring specifically here to the history of the acceptance of theories and not to the question of their truth.

We should note further that the tentative nature of scientific knowledge does not automatically mark that knowledge as relative in any sense that gives comfort to those opposed to the epistemic priority we give to scientific claims. Rather, scientists are rational in just the sense that CPR (the Commonsense Principle of Rationality) demands. In fact, the dynamic process by which scientists continuously revise what they are willing to endorse, and by which they examine their assumptions and their methods, is the heart of the strength of the sciences. Thus, despite the theory-bound nature of scientific knowledge, the self-critical (nay, reflexive?) process of scientific inquiry insures that whatever knowledge is claimed, it is the best available at that time. If that is relativism, so be it. But I am far more confident in the prospects of an ongoing process of self-appraisal that seeks to eliminate that which is unsupportable than with a dogmatism that relies on a misplaced sense of enlightenment and that makes timeless claims about a process (science) that occurs in time.

Ultimately, the aim of scientific inquiry is explanation. Using a theory, we explore a domain of objects, sorting out their various relations for the purpose of explaining what can't be explained otherwise by appeal to the activities of the objects in that domain. Why is a tabletop hard? To answer that question we have found that we need to appeal to a scientific theory that proposes that there is a domain of smaller objects that are held together by a series of forces and that it is because of the forces and objects in that microdomain that our phenomenological report of a hard table is possible. The aim of science is to help us understand the way the world appears to us, and it accomplishes this aim by constructing and testing theories that appeal to other features of the world that are not immediately obvious.[12]

There are other aspects of scientific knowledge that are essential to its vitality, but they need not be of concern. It is not necessary to explore the entire concept to set the ground for finding a counterpart concept. We need only concentrate on these two factors: (1) scientific knowledge is theory-bound; and (2) scientific knowledge is developed to explain the

12. This account of scientific explanation appears to endorse a form of scientific realism. The theory of explanation on which this view rests was developed by Wilfrid Sellars, and he was a scientific realist, meaning that he accepted the view that the ultimately real constituents of the world are the theoretical entities posited by our best confirmed theories. I accept the structure of Sellars' theory of explanation and replace Sellarsian scientific realism with Sicilian Realism. Sicilian Realism rests on two points: (1) accepting the position that the entities postulated by the current set of accepted scientific theories are *all* real, the world being a very complicated place, and (2) rejecting the principle of reduction, by which the entities of one domain are said to be nothing other than compositions of the entities of the domain of this or that scientific theory, e.g., tables are nothing more than collections of molecules. Sicilian Realism is realism with a vengeance. This view has affinities with "perspectivism." For example, see Dupré 1996 (esp. 104-5), and Stout 1988.

way the world works, to have a fruitful starting point to investigate the nature of engineering knowledge.

Section 4. Engineering Knowledge

In *What Engineers Know and How They Know It*, Walter Vincenti picks up and develops a theme first introduced by Edwin Layton in his landmark paper, "Technology as Knowledge." Vincenti provides an account of engineering knowledge from the point of view of a practicing and deeply reflective engineer. Both Layton and Vincenti endorse the view that engineering knowledge, and technological knowledge in general, constitute a discrete form of knowledge, different from scientific knowledge. Endorsing the findings of A.R. Hall, Layton, in a later piece, his classic 1987 Society for the History of Technology Presidential Address, "Through the Looking Glass or News from Lake Mirror Image," claims that "technological knowledge is knowledge of how to do or make things, whereas the basic sciences have a more general form of knowing" (Layton 1987, 603). Vincenti echoes this, invoking Gilbert Ryle's famous distinction between knowing how (technology) and knowing that (science).

Both Layton and Vincenti are concerned to defend the view that while both science and technology may borrow from or rely on each other in various ways, they constitute two distinct forms of knowledge, since they aim at differing ends; science explains and technology/engineering creates artifices. Vincenti puts it this way: "technology, though it may *apply* science, is not the same as or entirely *applied* science" (Vincenti 1988, 4). He defends this claim in part with an intriguing and highly suggestive proposal. As he sees it, if we start with the proposition that technology is applied science, then there is no possibility of considering that technology could involve an autonomous form of knowledge that could account for those technological achievements that are science-independent. This, as it turns out, is a variant of the view proposed in Chapter 2. Another way to make Vincenti's point is to observe that the sciences are deeply embedded in a technological infrastructure. But from this it does not necessarily follow that science is merely applied technology. From the fact that each activity has occasion to rely on the other, it does not necessarily follow that one is a subset of the other. That being the case, what can we say about the distinctive nature of engineering knowledge as a specific form of technological knowledge?

Starting from a wonderfully succinct definition of "engineering" by G.F.C. Rogers that is highly reminiscent of Mesthene's definition of "technology," Vincenti identifies three main components of engineering and then concentrates on the notion of design. According to Rogers (as quoted in Vincenti and augmented somewhat by me):

> Engineering refers to the practice of organizing the design and construction [and I (Vincenti) would add operation] of any artifice

> which transforms the physical [and, I (Pitt) would add, social] world around us to meet some recognized need. (Vincenti 1988, 6)

One of the commendable aspects of Rogers' definition is his characterization of engineering as a practice. That is, engineering, like science, is an *activity* with specific objectives. Mesthene's definition of "technology" was "the organization of knowledge for the achieving of practical purposes." By a series of substitutions we see that, appropriately enough, *engineering knowledge concerns the design, construction, and operation of artifices for the purpose of manipulating the human environment.* Concentrating on design, Vincenti further narrows his focus to the topic of "design knowledge." Here it is worth quoting Vincenti's description of the design process at length. It is noteworthy for two reasons. First, it immediately introduces an important distinction between the design as a set of plans and the design process. Second, Vincenti's description of the design process reflects in important ways the input/output model of technology as humanity at work offered in Chapter 2.

> "Design," of course, denotes both the content of a set of plans (as in "the design for a new airplane") and the process by which those plans are produced. In the latter meaning, it typically involves tentative layout (or layouts) of the arrangement and dimensions of the artifice, checking of the candidate device by mathematical analysis or experimental test to see if it does the required job, *and modification when (as commonly happens at first) it does not. Such procedure usually requires several iterations before finally dimensioned plans can be released for production. Events in the doing are also more complicated than such a brief outline suggests. Numerous difficult trade-offs may be required, calling for decisions on the basis of incomplete or uncertain knowledge. If available knowledge is inadequate, special research may have to be undertaken.* (Vincenti 1988, 7—emphasis added)

The process Vincenti describes is "task-specific" and essentially characterized by trial and error. But that still doesn't tell us in what design knowledge consists. That is because to capture the exact kinds of knowledge required, Vincenti must invoke a detailed model which breaks that process up into both vertical and horizontal components thereby allowing the precise identification of what is needed when and where in the total design process. The schema is proposed for what Vincenti calls normal design, as opposed to radical design[13]. Normal design has five divisions, beginning with the crucial aspect of any problem-solving process, the identification of the problem. An aeronautical engineer, Vincenti draws from his own discipline for appropriate examples, but the schema is general enough to encompass a large number of design processes, e.g., the design of an architectural project, including siting of the building, electrical systems, plumbing, etc., or the design of a space-based orbiting telescope:

(1) project definition—translation of some usually ill-defined military or commercial requirement into a concrete technical problem for level 2;

(2) overall design—layout of arrangement and proportions of the airplane to meet project definition;

(3) major-component design—division of project into wing design, fuselage design, landing-gear design, electrical-system design, etc;

(4) subdivision of areas of component design from level 3 according to engineering discipline required (e.g., aerodynamic wing design, structural wing design, mechanical wing design);

(5) further division of categories in level 4 into highly specific problems (e.g., aerodynamic wing design into problems of platform, airfoil section, and high-life devices). (Vincenti 1988, 9)

The process Vincenti outlines appears simple enough from the outside. One defines the problem, breaks it into components, and subdivides the areas by problem and specialty required as needed. What is not obvious at first glance is the way in which the levels interact. On reflection one can see that what happens at level 3 will have ramifications for the overall design and vice versa. In short, any design project must allow for a good deal of give-and-take throughout the process. As Vincenti puts it, "Such successive division resolves the airplane problem into smaller manageable subproblems, each of which can be attacked in semi-isolation. *The complete design process then goes on iteratively, up and down and horizontally through the hierarchy*" (Vincenti 1988, 9, emphasis added). If, by way of example, we apply this way of thinking to an architectural problem, we

13. Following E. Constant in his *The Origins of the Turbojet Revolution.*

can easily fill in the blanks when faced with the question of what kind of a building to design (level 1), e.g., specific or multipurpose, as opposed to the kinds of bathroom fixtures to have (level 4), although the one will ultimately bear on the other.

At this point we can pause and take a first pass at comparing scientific and engineering knowledge. The characterization of scientific knowledge in Section 3 as theory-bound and aiming at explanation appears to be in sharp contrast to the kind of knowledge Vincenti seeks. Engineering knowledge is task-specific and aims at the production of an artifact to serve a predetermined purpose.

There is one more important difference between the two forms of knowledge that Vincenti's account of engineering knowledge helps highlight. With engineering cast as a problem-solving activity (not in itself a characteristic that distinguishes it from other activities such as science or even philosophy), the manner in which engineers solve their problems does have a distinctive aspect. The solution to specific *kinds* of problems ends up catalogued and on the shelf in the form of reference works that can be employed across engineering areas. For example, measuring material stress has been systematized to a great extent. Depending on the material, how to do it can be found in an appropriate book, which makes engineering knowledge transportable. This gives rise to the idea that much engineering is "cookbook engineering." What is forgotten in this caricature is that another part of the necessary knowledge is knowing what book to look for. That is one special kind of knowledge engineers bring to their form of problem-solving.

Section 5. Philosophical Problems

In contrast, scientific knowledge is not clearly "transportable" across fields in the same way as engineering knowledge. One crucial obstacle presents itself: the problem of incommensurability.

The problem of incommensurability is a philosophical problem that came to the forefront in large part with Kuhn's characterization of the nature of scientific change. For Kuhn, change in science occurs through paradigm replacement. On his view, incommensurability applies primarily across paradigms. A paradigm for Kuhn is many things.[14] For the purposes of this discussion, however, let us consider it a complete system of thought including methodological rules, metaphysical assumptions, practices, and linguistic conventions. Two paradigms are incommensurable, it is alleged, because claims made in different paradigms cannot be compared so as to determine which claim from which paradigm is true.

14. C.f. Margaret Masterman in *Criticism and the Growth of Knowledge* (Lakatos and Musgrave 1970, 61).

For this view to be plausible, a particular theory of meaning must be assumed and a very dubious metalinguistic assumption must be activated. First, let us look at the theory of meaning. Basically, the theory of meaning behind the assumption of incommensurability presumes that expressions receive their meaning contextually, within systems, i.e., paradigms, governed by unique sets of rules. This by itself is not so troublesome. The difficult part comes through the metalinguistic assumption that there is no point of view common to both paradigms from which to compare claims from different paradigms. Such a common neutral point of view is necessary, it is argued by its proponents, since the meanings of expressions are governed by the rules of the paradigm. If we shift an expression from one paradigm to another, its meaning will change since it will be determined according to different rules.

Among other difficult problems to sort through here is the apparently unjustified twofold assumption that there is *one* fundamental theory of meaning that applies to all paradigms, i.e., the meanings of expressions within any particular paradigm are determined by the *rules* of the paradigm, but by contrast there is no single theory of meaning that allows for comparison of expressions across paradigms. But, if we can assert that all paradigms provide meanings for the terms that occur in that paradigm through the specification of rules, then why can we not, in the same metalanguage in which we pronounce this dictum, create another paradigm, a meta-paradigm, with the express purpose of allowing for the comparison of expressions? It is, for example, not at all obvious that the ways by which terms are made meaningful is through the specification of rules. That is, however, the account we are considering, and it is the source of Kuhn's problem of incommensurability. That much has been stipulated through Kuhn's account of a paradigm. But, unless something further prohibits us from doing so, surely we can say something like this for the purpose of comparing two expressions, each drawn from a different paradigm: if the result of applying those expressions in the metalanguage, according to the rules of the metalanguage, is the same *in the metalanguage, then for all accounts and purposes those two expressions mean the same thing.*

In short, if two expressions drawn from two different scientific theories yield the same result when transported into a third theory, then they can be said to make the same claim. But it is one thing to assert this, and another to show it. Furthermore, to do so would take us far afield. It is enough, however, to have identified the problem, to have seen where it came from, and to suggest a possible solution.

The solution is based on our account of engineering knowledge. If something formulated in the context of one paradigm can be used successfully in another, then deep philosophical problems about obscure theories of meaning appear to recede. To treat the problem of incommensurability this way is not to solve it as much as to ignore it. This too may not be a bad thing. There are many philosophical problems still around to which we no longer pay attention because they seem beside the point; for example, consider the pseudo-problem of how many angels can dance on the head of a pin. It is not clear that this problem was ever solved, but who cares?[15] Likewise for the problem of incommensurability. If the problem as stated was never solved, it appears not to matter. This lack of concern is a function of having *shifted our ground* from worrying about providing an abstract philosophical justification for something that only philosophers worry about to a pragmatic condition of success: consider the consequences of using this claim from this theory in this context.[16] If it solves our problem, then does it matter if we fail to have a philosophical justification for using it? To adopt this attitude is to reject the primary approach to philosophical analysis of science of the major part of the twentieth century, logical positivism, and to embrace pragmatism. This is a good thing to do, especially when we are concerned with technologies that have real-world effects.

---

15. It is equally unclear that this was ever a problem.

16. See Richard Rorty's introduction to *The Linguistic Turn* (1968, 1-39).

CHAPTER FOUR

Technological Explanation

In our discussion of scientific and technological knowledge we found not only that there were specific differences between the two types of knowledge, but that the urgency of certain philosophical questions concerning science did not carry over to technology. Unlike the case of scientific knowledge, there is no serious philosophical question about the transferability of technological knowledge. There *are* practical problems, which often arise due to contextual considerations. Thus, while the "know-how" to solve a particular problem may be available, the culture may not readily allow its use. (This is also true for scientific knowledge in some cases.) The key to dealing with the important philosophical questions concerning technology is not to assert that because incommensurability is a philosophical problem for science, it must also be so for technology, thereby flying in the face of what we know to be the case. That is, the real-world fact of the matter demands that when attending to technology we must accept that we will deal with different problems from those associated with science. Problems such as incommensurabilty, i.e., how to compare accounts drawn from different theories, are not philosophical problems of technology. That is because the only thing that counts when it comes to assessing technological knowledge is the outcome of applying the knowledge, not its justification. To assuage those who dislike dismissing philosophical problems I offer the following: if there is a technological counterpart to incommensurability, it will be found in a full-blown discussion of technology transfer between significantly different cultures. But, I suggest, in that context it becomes a practical problem.

This is not to argue that a similar case can or should be made for all counterpart questions. It is not yet at all clear if the outcome I suggested when examining the differences between technological and scientific knowledge will be repeated in the current topic: technological explanation. Following the earlier approach, I will first take a brief look at a philosophical account of explanation in science and then turn to technological explanation.

Section 1. Scientific Explanation

While it no longer holds sway over the topic as it once did, the Deductive-Nomological Theory of Explanation (DN) and its associated model remain the starting point for any contemporary discussion of explanation. The clearest presentation of that view, "Studies in the Logic of Explanation" by Carl Hempel and Paul Oppenheim (Hempel and Oppenheim 1948) remains the primary reference point and continues to exert a powerful pull on our intuitions about explanation. First, I take a look at the broad outline of DN and then several alternatives and expansions of it.

The DN model of explanation has a very simple structure, motivated by Hempel's understanding of what a scientific explanation is supposed to do: be an answer to a "Why?" question. What is to be explained is expressed in the form of a sentence. The thing to be explained can vary from a particular observation to an empirical generalization to a universal law. To accomplish an explanation you must deduce the sentence to be explained from a general law and a Statement of Initial Conditions. The assumed context is that the person seeking an explanation wants answers to questions like "Why did this happen?" or "Why does this look the way it does?"

A version of this model looks like this:

All *x* s [are] *y* s
This is an *x*
This is a *y*

The intuition behind the model is part of the conception of scientific knowledge discussed in Chapter 3, i.e., universality. Thus, for some event or generalization to be explained scientifically shows that under all circumstances of similar types, this is what will happen. There are also many other possible relations in which things can stand. The connecting term, "are," in the sample of the model is a stand-in for many different types of possible relations such as "do," "receive," "expect," etc. There are many different situations that call for explanations, and we want to make sure that the conditions governing the expression of the generalization that is essential to the explanation do not rule out reasonable candidates. Our first premise then can be not only:

All *x* s are *y* s

but also:

All *x* s do *y*

or All *x* s receive *y*

or All *x* s expect *y*, etc.

depending on the situation to be explained. The heart of the explanatory effort is carried by the act of including the event, situation, occurrence, fact, within the domain of a general law. That is, to *explain* something is to show that it is part of a universal phenomenon of the same sort and that this phenomenon is characterized by reference to general *causal* interactions.[1]

Since its publication in 1948, the model and its associated theory have been criticized, defended, elaborated, extended, revised, and made over in a variety of configurations.[2] Most of these efforts have been concerned to refine the theory by introducing the notion of a specific context for an explanation. These endeavors are also varied, some concentrating on the social context, others on the physical. Despite these efforts, the essential intuition remains the same: a *scientific* explanation accomplishes its goal by causally linking the phenomenon to be explained to a universal generalization whose justification has already been established.

Section 2. Technological Explanation

Since the heart of the DN model concerns the role of scientific laws, a DN counterpart for technological explanation must rely on some notion of a technological law. Are there technological laws?

Let us try to answer the question by asking another: What would technological laws be laws about? Scientific laws are laws about the structure of the universe and the relations among its parts. If we are correct, and technology is humanity at work, then technological laws would be about people and their relations. Technological laws would be the result of research done in the social sciences, since they have as their domain of inquiry human beings and their interactions. Surely this result will not rest easy with the intuitions of many. But perhaps I can build a case in its favor.

The case in favor of technological laws as social science laws comes in two parts. First, I rehearse some of the reasons for using the broad-based conception of technology that we have adopted. Second, I distinguish between technological explanations and technical explanations, where technical explanations are concerned with particular technological artifacts.

The basic reason for adopting the definition of technology as humanity at work was to make it possible to include the full range of human artifacts within our conception of the tools we develop for achieving our goals.

1. The adequacy of the intuition of explaining something by showing it to be part of a universal phenomenon can be tested by noting that parents rarely accept "But everyone is doing it" as either an explanation or a justification.

2. See Pitt 1988.

That means that social tools like governments, bureaucracies, and legal systems all count as technology. And this is as it should be, for these social structures are in fact tools for our use. It should come as no surprise, then, that on this account, technological laws would turn out to be laws about people and their interactions.

Furthermore, this does not mean that such laws would necessarily not be concerned with technological artifacts such as machines and large-scale systems such as electric power networks. For it seems well within the realm of possibility that there could be different kinds of laws describing the behavior and relations between individuals in society, some of them concerned with purely interpersonal relations, some concerning the effects of social institutions on interpersonal relations, and some concerned with the behavior of individuals when involved with various devices and differing environments. Thus there could be laws describing the behavior of people who are given hammers, e.g., whenever a person is given a hammer he or she tests the balance of the hammer before doing anything else. At the moment this seems more like a crude empirical generalization than a law, but it suggests the sort of thing that might be developed and refined.

The reason this discussion of social/technological laws is so imprecise is that social scientists do not agree on the correctness of any of the variously proposed laws alleged to govern institutions, societies, or individuals and their interactions.[3] This does not rule out the development of laws about which they might agree in the future. Furthermore, we should acknowledge that there are plenty of candidates for laws out there. In fact this may be part of the problem. Instead of taking one or two fairly well-developed theories of human behavior and testing them in detail, each social scientist seems inclined to ignore what others have done and to build his or her own account. Thus there are many potential candidates for laws in the social realm, but there is little agreement on which ones are the most promising and on how to decide among them. This leaves us in a rather unhappy situation.[4]

This situation may not be as frustrating as it sounds at first. Perhaps it merely speaks to the difficulty of the enterprise. Human beings are diffi-

---

3. See Mills 1959, and Diesing 1991 (esp. chaps. 6 and 7 for a fascinating account of the tensions between macrosociologists [the Merton school] and microsociologists of science [Bloor, Collins, Pickering, Latour, and others]).

4. There is another way to approach the issue of explicating technological explanations. This would be to abandon the effort to develop counterpart concepts to DN and to look instead at a totally different account of explanation. The account I have in mind is a pragmatic theory of explanation that places the emphasis on contextualizing events in an historical domain and then to consider the long term results of the developments under analysis. See Pitt 1994.

cult to tie down in terms of behavioral generalizations. Part of the problem is that we keep insisting on changing our environments in unpredictable ways, so that even if it is possible to formulate a generalization, it is difficult to confirm it because the situation to which it was supposed to apply has changed yet again. There is also another way to approach the issue of technological explanation; it involves developing a counterpart concept to DN by looking instead at an account of explanation that does not invoke laws.

The difficulty of formulating laws about people and their use of tools, simple or complex, mechanical or social, is the problem if we seek a DN counterpart account of technological explanation. For unless there are laws, no counterpart concept to that of a scientific explanation will be forthcoming. A second problem concerns the fact that in many cases, what people want when asking for a technological explanation is one of several accounts of the ways created and manufactured things work. That is, often people seek an account of the *technical details* when looking for an explanation. So, instead of technological explanations, maybe we should turn our attention to technical explanations.

Section 3. Technical Explanation

When we turn to technical explanations, we appear to face a different set of intuitions. Unlike scientific explanations, where the objective is to tie a particular occurrence to a more general set of universal truths, technical explanations seek an understanding of particular events in terms of other specifics. Ultimately, an accounting of the why and the wherefore may require an appeal to scientifically established universal claims, but the primary demand in a technical explanation is for specifics. Thus, if we want to know why a bridge fell down or a nuclear reactor exploded, we begin by asking why the Tacoma Bridge collapsed or why the reactor at Chernobyl blew up, not why bridges in general fall down. The appropriate answers to our very specific questions will not come from appeals to generalizations such as "gravity," in the one case, or "because it went critical" (with an accounting of all the attendant laws governing nuclear fission), in the case of the other.

There are basically three different types of technical cases that call for explanation:

(1) an artifact has failed to perform to expectations;

(2) someone wants to know how an artifact does the things it does or achieves the results it does; and

(3) unintended consequences.

That is, "Why doesn't it work?" "Why *does* (meaning "how does") it work?" and "Why did *that* happen!" Like the questions that motivate scientific explanations, all of these resemble "Why?" questions (Bromberger 1992, 75-100). But, as we shall see, they will not be answered satisfactorily in the same way.

But before beginning our discussion here, I need to consider one objection to our proposed taxonomy of explanatory questions. One might argue that unintended consequences can be seen to fall naturally into the first category above. Thus unintended consequences would constitute a special kind of failure. But they can also be seen as a special kind of success. For example, the birth control pill not only gave women control over the number of children they would bear and when they would bear them, but it also helped liberate them from the roles of mother and homemaker, freeing them to take on (or not) both careers and motherhood at their own choice. Since, as this example shows, the unexpected may not always be reducible to failure, unintended consequences rightly deserve their own treatment.

Equipped with this taxonomy of technical situations calling for explanation, we can now ask: What is the appropriate form of explanation for each? Or is there only one basic form of technical explanation?

Neither technological failure, success, nor unintended consequences can be accounted for in universal terms. The Tacoma Bridge disaster did not occur because of a single cause; there were many factors involved. Some of the possible variables include meteorological conditions, geological factors, material stress, poor design, poor material quality, failure to meet specifications, and lack of knowledge about the interplay of all the above (Anonymous 1983; Page 1983). Likewise for Chernobyl, but add to the earlier list "human error" (Chernousenko 1991; Shcherbak 1989). Despite the fact that the design of the reactor at Chernobyl was in use at a number of other sites in the Soviet Union, it was the reactor at Chernobyl that failed, and it is that failure that requires an explanation. A complete explanation for each of these failures will look like a list of specific contributing factors, not a generalization over them, because the failure of a bridge or the blowup of a reactor is specific to place and time. Nevertheless, there is an implicit appeal to the power of invoking something like a law when opponents to some technology seize on a failure of that technology and generalize over it in an attempt to block further development. Thus, Three Mile Island and Chernobyl had accidents, hence all nuclear reactors will have accidents. But human participation is also an active component, and that participation varies depending on innumerable factors. Thus I would argue that in principle there can be no appeal to a single general technological law in the spirit of DN that will explain what went wrong.

Now it might be argued that exactly the same problem of specificity occurs when we attempt a *scientific* explanation of any specific event.

Don't earthquakes, volcano eruptions, etc., all need multiple-factor specific explanations, with no pretense to a single law of, say, volcanology?[5] Yes and no. Yes, a scientific explanation of a natural occurrence may need to appeal to multiple factors. The difference is that while each of these events requires what will ultimately be a very complex explanation, on DN the structure of the full and final explanation remains fundamentally the same: a deduction. It is a complicated deduction with lots of nested arguments and deductions within it, but it remains a deduction nonetheless. Furthermore, the reason that such a deduction is possible is that while there is no one single law of say, volcanology, there is, or could be, nay, must be a *theory* of volcanology, and it is within the framework of this theory that each of the specific explanations yielding the final complex explanation is forthcoming. Remember that science is theory-bound. This is but another example of that phenomenon. On the other hand, something requiring a technical explanation does not also require that each of the various explanations that contribute to our understanding of what happened fall under a systemic theory of the whole.

To see why technical explanations must be concerned with the specifics, compare, for example, the case of the Tacoma Bridge collapse and the recent collapse of an interstate highway bridge in Connecticut. In the Tacoma Bridge example, a combination of lack of knowledge about the effects of winds on suspension bridges and an inadequate design including bad placement contributed to the failure. For the interstate highway bridge, poor-quality concrete was a major contributing factor. The contractors skimped on the concrete in order to save money and increase their profit margin (Berreby 1992; Petroski 1991; Wirkmann 1991). In crucial respects, very different factors play key roles. A major factor in the collapse of the interstate bridge was greed. Because of that greed, certain other consequences occurred. By way of parallel, we might say that the basic cause of the Tacoma bridge failure was human ignorance. But ignorance and greed are two completely different factors, which, depending on the circumstances, will have varying effects. When both are involved, it is not clear that we have a straight rule for deciding which is more important; nor is such a rule appropriate.

If we turn to explanations of why things work, the situations are generally simpler. One tried-and-true way to explain why artifacts work the way they do is to reason backwards from the end effect to its immediate cause and then to its *immediate* cause, etc. Thus, if we want to know why the light goes on when we turn the switch, we first look at how the switch activates the flow of current, then to the effect of the current on the lightbulb, then to

5. I owe this example to my colleague, Harlan Miller.

the source of the current to that switch through the house wiring, back to the main electrical panel, to the line feeding the house, etc. Or if we want to know what causes the car to move, we trace the line of effects and causes back from the wheels to the axle to the drive shaft, and so on.

But, it might be objected, doesn't this beg the question? Isn't the real issue in explaining why the light turned on an account of the nature of electricity and, for the motion of the car, the scientific principles behind combustion? Not so; consider the following. Picture, if you will, a small dam with a mill race directed into a large waterwheel connected to some mechanism in the building on the shore of the river (the mill). You want to know what makes the waterwheel turn. I show how the force of the water hitting the sections of the waterwheel pushes the wheel down, and that is why it turns. I can take your hand and give you an exercise in pushing down on a small wheel and making it turn, then suggest that the water acts like my hand. There is no need here to appeal to physics and provide the appropriate sophisticated account of "force." In point of fact, if such an account as F = MA were provided, it would do little to explain why *this* waterwheel turns. For one thing, there is a big jump from the abstract

equation's definition of "force" to the identification of the water from the mill race as the force in question. Physics does not identify the things that act as forces in the world, nor should it, since what is a force in one situation may not be one in another, e.g., water in a bowl does not act as a causally active force.

Thus the basic idea behind an explanation of why some technological artifact works the way it does is to work backwards from the identified effect to be explained to its immediate cause, and to *its* cause, etc. One final objection to this account needs to be considered. By suggesting that we work our way backwards, haven't we opened the door to an infinite regress? Isn't it going to turn out on this approach that the ultimate cause of the car moving is the Big Bang? This need not be the result. The primary thing to consider here is that the request for an explanation is not made in a vacuum. When we speak of technical explanations, we are referring to explanations of specific effects of specific artifacts or systems of artifacts. Thus we want to know why *this* car moves or *this* waterwheel turns. Questions about "the" cause of this car or "the" cause of this waterwheel should be seen as asking for the identification of a different set of causal antecedents. In short, questions about the workings of this artifact come to closure when questions about the cause of this artifact are raised.

Turning now to explanations of unintended consequences of a technology, the glib answer here to questions such as:

How could we fail to see that this could happen?

is

We are not omniscient.

But, glib as it may appear, there is a ring of truth here. The failure to anticipate all possible consequences of introducing a new type of artifact or a new way of doing things is often not merely due to not having all the available knowledge to make accurate predictions. Sometimes lack of knowledge *is* the cause of an unexpected result of developing and/or introducing a new technology. Surely when Henry Ford developed mass-production techniques for the construction of automobiles, he did not foresee the rise of suburbs and commuter traffic. But just as often, the unexpected impact of an invention is the result of human ingenuity. In this case, consider the rise of nuclear medicine as a result of the Manhattan Project. Unintended consequences must not be seen necessarily as a failure on the part of planners and designers. Rather, given the ability of human beings to take new ideas and objects and adapt them to their own needs and goals, the occurrence of unintended consequences is a basic feature of the world of technology, the world of humanity at work.

It is at just this point that we can extend our discussion beyond questions we have been discussing, demonstrating the fertility of approaching questions about technology from this perspective. Taking the line advocated earlier on unintended consequences, for example, opens the door to a discussion of technological determinism. By not only admitting that there are unintended consequences of technological development due to human ignorance, but by also suggesting that the one thing humans cannot anticipate fully is the way other humans will react to something new, we are implicitly suggesting that there is no inevitable causal line from a new device to its impact on society. That leaves a lot of room to maneuver with respect to the question of whether technological innovations restrict human actions or increase their options. And by making the epistemological questions about our ability to know what humans will do the central feature of this discussion, we make the subsequent social policy questions far more complicated and far more interesting.

As we have seen, it does not appear that we have a parallel counterpart concept of technological explanation for scientific explanation. If universal laws are needed in order to dress technological explanations in scientific clothing and hence give them a "real" explanation, then since those laws must be forthcoming from the social sciences, we had better not look for resolution of this approach soon.

On the other hand, if we direct our attention to technical explanations, we find ourselves concerned with the specifics of cause-and-effect relations, again avoiding appeals to universal relationships. Does this make what we produce, then, any less an explanation? It does not appear so. Fundamentally we are still concerned with causal relations, if only restricted to particular cases. We are still seeking to provide answers to "why" questions. This is consistent with general technological reasoning as we have attempted to capture it in our Commonsense Principle of Rationality (CPR). To learn from our experience and build that knowledge into our base in order to benefit from it on the next go around, then it must be knowledge of a cause-and-effect relation. Only then can the necessary adjustments take place.

This then suggests one further conclusion. The basic pattern of tracing individual causal relations seems to be the most fundamental way of answering "why" questions. From this it would seem that technical explanation is both empirically and logically prior to scientific explanation. And given the obvious linkage between technical explanation and CPR, it would seem that scientific rationality is at best a subset of technologically based everyday reason. In short, matters technological have logical priority over matters scientific; scientific activity is only one form of humanity at work.

This effort to explore, albeit all too briefly, some counterpart concepts between science and technology leads to the conclusion that the generally assumed close linkage between these areas of human endeavor should be reexamined. Caution should be exercised when talking about "science and technology" as if they necessarily came as a package. More attention should also be paid to the social dimensions of technology, i.e., the various forms of social arrangements and institutions. From this it does not follow that we cannot pay attention to "things" or artifacts, the objects more popularly referred to as technology. It is not only possible but clearly central to thinking about things and their effects that we be able to ask and answer specific questions about the way things work.

The distinctions and rethinking outlined here are not intended to be mere exercises in wordplay. When we leave the people who make the machines and who make the decisions to build machines and to deploy them in systems in specific environments out of the picture, we do our innovations, inventions, and ourselves a vast disservice. Technology writ large does not come from nowhere, capture and bind us to its will, and then march inexorably into the future, damning humanity to a mindless existence. *We* build the support systems, the factories, the global networks. We learn from them, use them for better and for worse. They constitute our most obvious achievements, from plowed and planted fields to space shuttles. We should not forget, as we seek to understand what we have built, that *we* have done all this and that our role cannot be ignored without the risk of failing truly to understand what is happening to us and what we are doing to ourselves. Let us look at an example.

Section 4. An Example: The Case of the Hubble Space Telescope

Responsible social criticism of a technology requires first that we understand the technology in question, and second that we use reliable and trustworthy methods to determine the accuracy of factual claims that are either in opposition to or in defense of the use of that technology. As philosophers of technology, that puts us in the methodology business. One crucial engineering methodology is design. To be concerned about methodology is, among other things, to be engaged in attempting to develop criteria for evaluating empirical claims made about the efficiency of the design process. In short, it is not enough to understand technology, however contextualized. Rather, we need to know how that knowledge is constrained by the methods, assumptions, and values we or others bring to the investigation.

In that spirit, and in an attempt to see how some of the abstract considerations elaborated earlier work out, in this section we look at three different claims about the design process: Walter Vincenti's reiterative model, discussed briefly in Chapter 3, Louis Bucciarelli's socially situated model, and the model MT proposed in Chapter 2. Our objective is to see how well they assist us in unraveling the complexities of real-world activities. After presenting the three models, some less fully than others, we turn our attention to a case study that, while only partially developed, has the merit of having at its heart a flawed project: the Hubble Space Optical Telescope (HST), a product that embodied an engineering/manufacturing mistake. Here we have an example of a flawed design process. We will see that neither Vincenti's reiterative model nor Bucciarelli's claims about the social nature of the design process bear up when examined in light of what happened to the Hubble. What about MT? Well, as far as I can see, it comes closest to an account of the technological process that can accommodate what went wrong as well as handle a process that works out without problems.

Why is it so important to accomplish this goal, i.e., explain what went wrong? More is at stake than simply finding out why a project didn't meet the constraints of some theoretical model or other. It has, rather, to do with the very question of assessing the adequacy of theoretical models and in turn, among other things, exposing the ideological element in our models. In this case, the ideology is what I call the Myth of the Engineer, in which engineers are portrayed as the paradigms (*pace* Kuhn) of rational and project-oriented problem-solvers.

In labeling this ideology the Myth of the Engineer, I do not aim to offend. I seek to understand how things happen by applying an approach we have been developing in science and technology studies to understand how science works, Putnam's miracle warts and all.[6] In its infancy, the philosophy of science started with the assumption that science is our best means for achieving knowledge about the natural world, and philosophers of science set about attempting a justification for this claim. It just well may be true that science is our best knowledge-generating machine, but that needs to be discovered, not assumed. The fully contextualized, historical, and socially informed philosophy of science that is now emerging justifies the early assumption of the epistemic superiority of science, but it also exposes the mess that is real daily scientific activity and it thereby provides a deeper appreciation of the genuine accomplishments of scientists when they do emerge. More important, it reveals the extent to which the process of science is imbedded in technological infrastructures. In our current case the connection is obvious: the astronomical discoveries to be made require a telescope. But this is not just any telescope, it is a space telescope that requires a means to get it into space. Further, the support package needed to make the telescope functional once in space consists of

6. Putnam claims that "the typical realist argument against idealism is that it makes the success of science a miracle" (Putnam 1978, 18).

a set of additional instruments such as gyroscopes to keep the telescope steady and accurately pointed. And as we continue, we will discover the extent to which the success of whatever scientific missions are associated with the Hubble is tied to the success of its technological components and the ability of the support team to use additional technological means to correct failures.

There is one more reason for looking at a project gone wrong rather than concentrating on engineering success stories. In epistemology, one of the criteria of adequacy for any theory of knowledge is that it be able to account for how we can make false knowledge claims as well as why knowledge claims are warranted. The process by which technological projects are accomplished is, as I have been arguing, analogous in many ways to the knowledge-producing process of science as well as being disanalogous in other ways. To the extent that engineers are successful in producing products and in codifying the means for generating equal success in other projects, they develop something analogous to scientific theories and protocols. If that is correct, then philosophers of technology should be able to lay down criteria for determining if a project is a success *or not.* I will be attempting to show that Vincenti and Bucciarelli do a very nice job of producing models or theories that account for some aspects of successful design processes, but not for ones that fail. And for those that fail we need something stronger than falling back on the position that if the project failed, the participants weren't very good engineers. I am not claiming that this is what Vincenti and Bucciarelli do, but it is what their models make possible. Attacking the professionalism of the participants shifts the blame away from the inadequacy of our models in an *ad hoc* manner. This move is unhelpful since, I assume, we first seek to understand what happened, not to assess blame. Assessing blame is the aim of the social critics. This can be legitimate criticism, but it should come later, after we know what actually happened. More to the point, if we *show* that a project failed because it didn't follow our idealized models, and then we turn around and *explain* the failure by blaming the participants for being defective in some way, we have in fact bought into the Myth of the Engineer. We are in effect saying that no well-trained engineer would have allowed that to happen, and since it did happen, then these folks can't be good engineers, hence this failure does not count against this particular model and its idealized account of how engineers ought to behave. That move, I suggest, is just as flawed a methodology as any other ideological presentation of a position. (More on ideology in Chapter 5.)

Part of what I have been arguing is that before we engage in assessing blame and in social criticism, we need to know what really is or was going on. That is accomplished at Stage One of a two-stage process. Properly done, social criticism is a Stage Two activity. Stage One activities are designed to uncover the facts of the situation.[7] While good social criticism requires good detective work at Stage One, we also need to know what to look for in order to move to Stage Two. I propose that *we look at who made what decisions and why.* Only by concentrating on the decision-making dimension of technological development can we proceed to responsible Stage Two work. (Warning: do not confuse the talk of Stage One and Stage Two here with the different components of the model MT discussed in Chapter 2. The point of making the Stage One/Stage Two distinction here is to locate the appropriate place for social criticism, not to schematize the technological process, which is the objective of MT.)

A couple of preliminary notes before proceeding to this discussion of the three models of the design process. First, I have located my discussion of technological issues in the context of engineering. The second point concerns my focus on design. If there is a similarity between technology and science, then the design process is to technology what the scientific method is to science. But just as there is no scientific method *simpliciter,* there is no design process pure and simple. There are many, and they are the methodological counterparts to methods of experimental design and testing in the sciences. In what follows I will be looking at models of the design process to determine whether or not they stand up to what engineers actually do—are they explanatory, and if so, in what sense?

As we saw, according to Vincenti:

> Design, apart from being normal or radical, is also multilevel and hierarchical. Interacting levels of design exist, depending on the nature of the immediate design task, the identity of some component of the device, or the engineering discipline required. (Vincenti 1988, 9)

His account of the design process listed five stages: project definition, overall design, major-component design, subdivision of areas of component design from level 3 according to engineering discipline required, and further division of categories in level 4 into highly specific problems. "The

7. Before this claim gets laughed out of court by the postmodernists, let it be noted that, cynicism about knowledge aside, there is a world "out there," and things do happen in it. No postmodernist is going to deny the reality of gravity and walk out of a twelve-story window expecting not to fall to serious if not terminal injury. The extent to which we are able to discover what actually happened in the world and relay that information in as neutral a means as possible is going to vary, depending on many factors, including the reliability of our methods, the accuracy of the actors involved, etc. However conceived, the impulse to explain how something came to pass must assume that there are facts to be uncovered. If this is not the case, then postmodern appeals to the rhetoric of science or to the social are self-refuting because they must assume the rhetoric or the social is real if it is to have the explanatory value they wish.

complete design process then goes on iteratively, up and down and horizontally throughout the hierarchy" (Vincenti 1988, 9).

Despite the fact that Vincenti couches his analysis in the framework of airplane design, as noted earlier, the general process seems generalizable to any other design project, be it architectural, landscaping, curriculum design, etc. Further, it allows for movement back and forth between the levels and interactive readjustment. What it leaves out in the form in which it is expressed are the people and the social environment in which this all takes place. Granted, Vincenti acknowledges that decisions are made and that there is give-and-take, but he does not identify an overt role for those who make the decisions. This turns out to be crucial, as we shall see. And this distancing of the model from the messy social relations that develop whenever actors are included in the discussion is what Bucciarelli attempts to correct.

According to Bucciarelli:

> There is no science of design process in the way the participants understand that term. This is not to say that the process is irrational, that a story can't be developed and told that makes sense, or that one cannot, on the basis of this story, infer improvements in the process. It *is* to claim that to be "scientific" about the study of design process one must admit the possibility that the object—as either physical principle or economic necessity—is only part of the picture, and a very fuzzy part at that. If we want to understand the design process, we must remain sensitive to the full breadth and depth of social context and historical setting. (Bucciarelli 1994, 18)

Bucciarelli's initial contribution here is to emphasize the social and the historical and, most important, to recognize that the object is only part of the picture. Unfortunately, Bucciarelli continues a little further on to strengthen the claims about the social in ways that render his project suspect:

> My working hypothesis is that the process is not autonomous, that there is more to it that the dressing up of a scientific principle, more than the hidden-handed evolution of optimum technique to meet human needs, and more than the playing out of the bureaucratic "interests" of participants seeking power, security, or prestige. In the affirmative this hypothesis takes the form: *Designing is a social process.* (Bucciarelli 1994, 20)

So far, so good; Bucciarelli is taking the move to the social and distancing himself, so it seems, from the more radical social constructivists in science studies. But there is more:

> Executive mandate, scientific law, marketplace needs—all are ingredients of the design process, *but more fundamental are the norms and practices of the subculture of the firm where the object*

> *serves as icon*. . . .In the simplest terms, design is the intersection of different object worlds. No one dictates the form of the artifact. Hence design is best seen as a social process of negotiation and consensus, a consensus somewhat awkwardly expressed in the final product. (Bucciarelli 1994, 20-21, emphasis added)

As we shall see, the claims about the lack of outside interference on the form of the artifact are a bit too strong; likewise, the overwhelming emphasis on the norms of the firm; and it is perhaps also too strong to claim that "No one dictates the form of the artifact," for, especially where government contracts are involved, there are specifications to be met. Thus, while the exact form of the artifact may not be dictated, its performance measures are. All this aside, however, Bucciarelli makes an important contribution to our understanding of the design process through his emphasis on the internal workings of the firm and the notion of compromise that comes out of a realistic, i.e., accurate, account of the negotiation process.

According to the model in Chapter 2, technological processes, which essentially involve design processes, consist of two levels of input/output transformation and a third component of assessment, which also operates at both levels. The emphasis of the model is on decision-making. It stresses continual integration of new knowledge into the knowledge base of decision-makers and continual reassessment of that knowledge base and values in the light of new knowledge. In that sense it differs markedly from Vincenti's and Bucciarelli's models by focusing on individuals, not abstract process nor the social broadly construed.

We now have our three models. Let us turn to the problem at hand. How did NASA manage to launch a defective telescope? Or, not to appear to put the blame on NASA, how did a defective telescope come to be launched?

We can divide the history of the HST into four stages: planning, manufacture, launch and stabilization, repair. The initial idea for an orbiting telescope is credited to Lyman Spitzer in a report he did for RAND in 1946 (Smith 1989, 30). I will not spend time on the period between that initial proposal and the final funding authorization by Congress. It is a complicated and fascinating story elegantly told by Robert Smith in his *The Space Telescope*. Nor will I discuss what was involved in launching, stabilizing, and repairing the HST.[8] Instead, we will concentrate on some of the factors involved in its manufacture, those factors which had the result of sending a flawed telescope into space.

The Hubble experienced a number of difficulties, from the initial difficulty in getting the lens cover to open, to the failure of the solar panels to unfold completely, to the chronic instability of the telescope once in orbit,

---

8. For a fascinating insider's account of the conflicts among some of the major groups responsible for launching the Hubble and bringing it on line once in orbit, see Chaisson 1994.

to the flawed main mirror. When we consider all of these problems, the eventual achievement is all the more impressive. This sense of achievement is reinforced when we also consider (1) the range of people and interests involved, from politicians to manufacturers to scientists to bureaucrats; (2) the range of scientific, technical, political, financial, and, not to be left out, bureaucratic constraints that had to be satisfied; (3) the sheer complexity of the problem of creating an orbiting observatory, i.e., that more was involved than building a telescope. There were coalitions to be created for each specific point along the march to completion. There were trade-offs to be made to satisfy competing interests, e.g., astronomers committed to earth-based telescopes versus those committed to space. Decisions were made based on factors irrelevant to the mission of the telescope that had a major impact on its design. Consider only the following three. First, while the launch vehicle for the orbiting observatory was initially imagined to be a rocket, once NASA made the decision to go with the Shuttle program, the Shuttle was then designated as the launch vehicle for the HST. It was deemed politically important to get the most mileage out of the Shuttle, and arguments for it included carrying the HST aloft. So the HST was held hostage to the Shuttle program. This meant that the size of the HST was restricted to what could fit inside the cargo bay of the Shuttle. Second, the *Challenger* disaster resulted in, among other things, delaying the launch of the HST for four years. Third, financial constraints were a major factor from the beginning. More to the point, cost overruns and an exhausted budget resulted in a failure to conduct tests that were mandated in the contract. Everywhere we look we find decisions being made that would affect the performance of the HST but that had little to do with the scientific objective of the mission.

To see how these sorts of decisions played out, we are going to concentrate on some of the factors leading up to the problem we are concerned with: the flawed main mirror. Some of the material here comes from the U.S. *Congressional Record* (November 16, 1993). Once the HST was finally launched, and after it was admitted that the blurry images could not be straightforwardly corrected, Congress ordered an investigation. This resulted in a report presented to Congress by William Colvin, inspector general of NASA. Several things are not in dispute:

(1) The primary mirror of the HST was flawed. It suffered from a spherical aberration.

(2) The flaw was the result of a manufacturing error and several failures on the part of management.

(3) The manufacturer of the mirror, following the investigation, agreed to repay the government $25 million. The willingness of the manufacturer, Perkin-Elmer, to agree to this settlement, suggests that something more than a mistake was involved.

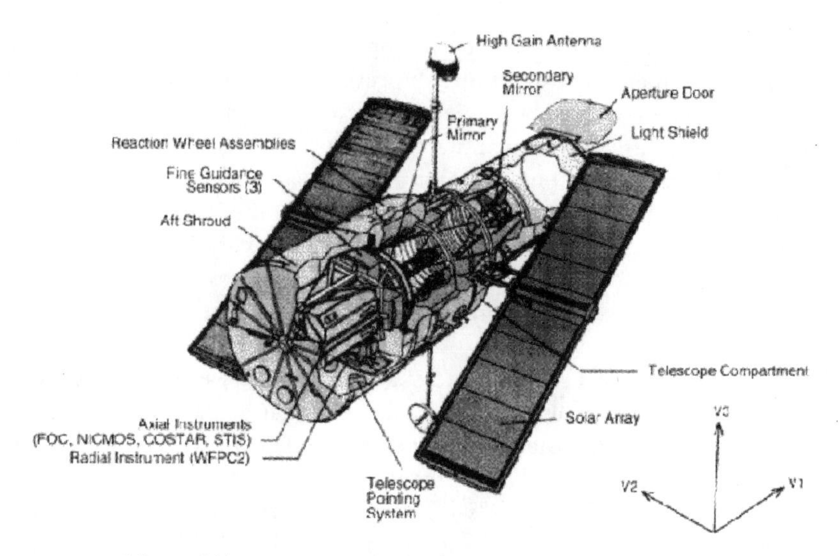

FIGURE 1 The Hubble Space Optical Telescope

Figure 1 gives us a diagrammatic look at the telescope in its space platform.

And Figure 2 shows an account of what a spherical aberration is, from the Space Science Institute at Johns Hopkins University—the scientists in charge of planning and executing the use of the Hubble.

The revelation that the telescope suffered from this specific problem and not from a failure to focus the primary and secondary mirrors was first proposed by scientists at the Space Science Institute (Chaisson 1994, Chapter 4). The NASA engineers and the engineers at Perkin-Elmer initially objected and sought to find the solution in adjusting the two mirrors for focus. After all efforts were exhausted, the engineers agreed to the diagnosis of a spherical aberration.

As noted, the result of the House of Representatives inquiry was a report from NASA following an extensive investigation. The investigation revealed the following six significant irregularities (and here I quote directly from the testimony of Bill Colvin[9]):

9. A word of warning here. If Chaisson (1994) is to be believed, there was continual bickering among the factions in NASA. Further, he claims that NASA made numerous blatant efforts to cover up what was actually happening with the Hubble. If he is correct, then maybe we should be careful in how much store we put in a NASA report. Consider the fact that no NASA officials are blamed in Colvin's testimony before Congress despite the fact that NASA officials and inspectors were involved in each stage of the manufacture of the HST. NASA did not simply award the contract and then sit back and wait for delivery. These worries aside, there is considerable value in Colvin's testimony for our analysis with respect to the emphasis he places on the decisions that were made.

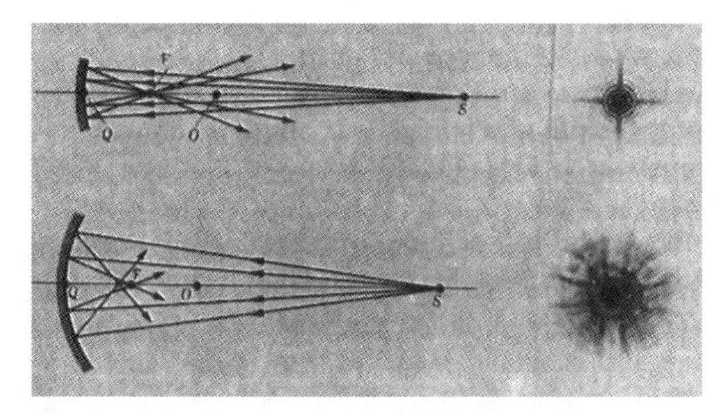

TOP: A perfectly shaped concave mirror brings all of its captured light to a single sharp point, called the focus. Here, in opticians' symbols, the center of the mirror is designated *Q*, the center of curvature of the mirror *O*. Light rays of a distant point source at *S* (considered for all practical purposes to be at infinity) then converge to a true focus at *F*, precisely halfway between *Q* and *O*. The result of such a perfectly focused mirror is a bright point with faint (diffractor) rings around it, such as the negative image of a star shown at the right.

BOTTOM: The blurriness of spherical aberration stems from an irregularity in the shape of the mirror that does not permit all the captured light to collect at a single point at *F*. Instead, a distorted spot of diffuse illumination appears with a bright point at its center, since different parts of the mirror focus light to positions in and around, but not all directly on, point *F*. Such an ugly image, shown at right to same scale as above, is inconsistent with that expected from a telescope that is merely unfocused or uncollimated (misaligned). Rather the excessive blurring can be understood only if a substantial amount of spherical aberration has been effectively ground into the mirror.

[Space Telescope Science Institute]

FIGURE 2

1. **Non-approved Reflective Null Corrector Washers.** "In the process of adjusting the spacings of the reflective null corrector for the HST Mirror, technicians discovered they could not move the field lens into the prescribed position. Instead of calling in the designer, the contractor inserted an ordinary washer under each of the bolts holding the field lens retainer to the adjustable plate. The insertion of these non-approved washers in an instrument whose precision is measured in tens of millionths of an inch required a nonconformance report. There is no written evidence that a report was generated" (p. 36).[10]

10. All of the following quotes are from Bill Colvin's testimony before Congress. Page numbers refer to U.S. Congress 1994.

A nonconformance report is a report describing how what was done deviated from the specifications of the government design. Why did the Perkin-Elmer engineers fail to file such a report? In an article in *Science*, the explanation rendered is that officials at both NASA and Perkin-Elmer "allowed themselves to be overwhelmed by the massive cost overruns and schedule slippages in other parts of the project. As a result they neglected the mirror work, which seemed to be relatively well, and failed to enforce their own quality assurance procedures" (Waldrop 1990, 1333).

This would seem to argue against both the Vincenti and Bucciarelli models. For not only do we not have evidence here of the reiterative process Vincenti prescribes, which should have intervened when the technicians couldn't move the mirror into the right position and should have called for the designer, but didn't, but also Bucciarelli's confidence in the firm's concern with the object seems to be missing as well. MT, on the other hand, focuses on the decision making. It deals with this situation by concentrating on why, for example, the decision was made not to call the designer back in. And if we turn to some of the other factors, such as cost overruns and schedule delays, these are externalities with a major impact on the final product. These externalities do not appear to have a way of being factored into the other two models. Vincenti emphasizes the structure of the design process. Bucciarelli stresses the social context, especially that of the firm. Neither apparently allows the externalities to play much of a role. My emphasis is on the decision making and the factors that bear on it, whether they be internal to the process or external. In the remaining five items in Colvin's report, each point comes back to the decision made by the individuals involved. The factors playing into those decisions varied; some of them were internal to the firm, some were not. However, only by directing attention to which people made what decisions with respect to the materials used and the objective to be obtained can we understand what happened. Let us return to the remaining five points in Colvin's testimony.

The second item Colvin relates is:

2. **Unexpected Results from Inverse Null Corrector.** "The inverse null corrector was used by Perkin-Elmer as a part of the reflective null assembly. It could be used as a check on the reflective null corrector's alignment and stability. It could show whether there was a gross flaw in the reflective null corrector as well as measure its stability. Since the inverse null emulated what the primary mirror surface should be when finished, it

should have produced an interferometric pattern with straight lines, or in other words a null condition. Instead, a pattern of wavy fringes was produced by the inverse null corrector at each testing cycle using the reflective null corrector. Although the inverse null corrector was producing results which did not agree with expectations, no nonconformance report was generated, and the designer was not consulted. While no qualitative analysis was performed, the inverse results were dismissed as being attributable to error in the null corrector's manufacture. As a result of this condition, Perkin-Elmer managers decided to make a design change which revised the usefulness of the null corrector from its initial utility as a "double check" on the health and stability of the reflective null. Through testimony we ascertained that Perkins-Elmer managers decided to use the inverse null corrector solely as a stability check" (p. 37).

The null corrector is an assembly of two mirrors and a lens that causes a laser beam to bounce off all parts of the curved mirror surface. This light is then collected and examined to determine the degree of uniformity of the curvature of the surface. The null corrector was employed in two different tests. First it was used to test the shape of the main mirror. In this first test, because the lens of the null corrector was incorrectly positioned, it actually contributed to allowing the flawed shape of the mirror to move on to the next stage. The lens was inserted a tenth of an inch too low. Because of that small discrepancy, and because of the size and measurements of the mirror, the test using the null corrector produced results that had the mirror meeting the required specifications, when in fact it did not.

The second test is the one Colvin is referring to. This time the results of the test were deemed flawed and the blame was placed on the null corrector.[11] This raises the unanswered question why the results of the first test were not then revisited and reassessed. If the null corrector was not performing correctly for the second test, we should not assume it performed correctly for the first test.

3. **Refractive Null and Reflective Null Tests Do Not Agree.** "Within a week of the first reserve null results, Perkin-Elmer received other unexplained testing results. A refractive null corrector was used by Perkin-Elmer in Wilton, CT, for initial

11. See Chaison 1994, 222–24.

grinding of the mirror blank to get the mirror surface shape roughly to the desired finished shape. The mirror was then transported to Danbury, CT, where testing was first conducted using the reflective null corrector. The test results from the first reflective null corrector test and last refractive null corrector test did not agree. Yet the expectation was the interferograms would be similar. In assessing the differences, Perkin-Elmer commented on the poor quality of the interferograms—caused by the rough surface of the mirror. They also asserted that some hand polishing of the outer edges of the mirror had occurred after the mirror left Wilton. No further quantitative analysis was performed" (p. 37).

4. **Recommended Gross Error Test Not Performed.** "At the conclusion of the polishing phase, a Perkin-Elmer Vice President and General Manager formally requested an internal review of the primary mirror certification by senior scientists employed by Perkin-Elmer. This group, called the technical advisory board, held a review of the mirror fabrication and test results. On May 21, 1981, they recommended to the Perkin-Elmer Vice President that an independent test be performed on the mirror. The recommendation stated that 'another test of the figure using an alternative method such as a Hartman test or a beam should be made. The purpose of the test would be to uncover some gross error such as an incorrect null corrector.' No such test was ever performed. If no such test was made, we can infer that the Perkins-Elmer Vice President decided not to make it. Why he decided that most probably has to do with the factors alluded to above, cost overruns and schedule delays" (pp. 37-38).

5. **Vertical Radius Test Anomaly.** "In May 1981, the most significant of the irregular events occurred when the refractive null corrector was used to measure the center of the curvature of the mirror. . . . the refractive null interferogram showed wavy lines—clear evidence of an error in one of the measuring devices and possibly the mirror. . . . the chance that two separate measuring devices, the inverse and refractive null correctors, would find matching errors and both be wrong and the reflective null right—is infinitesimal. . . . Yet Perkin-Elmer. . . failed to resolve the discrepancies in a quantitative way. . . . had Perkin-Elmer attempted to determine the source of the error, with the analytical and measurement tools in place at the time, they could have determined that the flaw was in the reflective

> null corrector in one or two days. Perkin-Elmer personnel assumed erroneously that there were large "as built" errors in both the refractive and inverse null correctors. They ignored the results of the intended sanity checks on the reflective null corrector. . . .the results of the last . . . test concerned Perkin-Elmer managers, but they did not disclose the results or their concerns outside of the optics fabrication group. . . . to our best determination, Perkin-Elmer did not share the discrepant results of the vertex test with NASA. The NASA Plant Representative . . . was provided a copy of the center portion of the vertex radius test interferogram. This cropped version did not disclose the curved fringes which would have indicated a problem with the test results" (pp. 38-39).

Here we are at a crucial point. Someone deliberately cut out the center of the interferogram in order not to reveal evidence of a problem. The decision to hide relevant evidence is at the heart of the resulting problem with HST, since as we shall see, other members of the Perkin-Elmer group were apparently left in the dark. Colvin's report continues: "Perkin-Elmer quality assurance personnel told us that they were present at daily meetings on the HST and they were never made aware of aberrant test results, nor were there any such discussions at the meetings." If this last point is true, then it suggests that communication within the company was flawed. It is not clear if there was competition or merely an effort not to look bad. It begins to sound as though upper management charged with overseeing manufacture of the mirror dug itself into a hole that further decisions simply deepened. I am not sure if this supports or undermines Bucciarelli's view.

Finally, there was rapid close-out of the Perkin-Elmer optical fabrication team. The program manager had eleven items he wanted to check out, but he was not given approval to do so. And in conclusion, "[T]he Head of Manufacturing Optical Analysis stated that he and his manager would have couched the vertex radius anomaly in terms of a 'need to recertify the reflective null corrector'" (p. 39). This is nothing more than an admission of intended obfuscation, which tends to support my earlier conclusion.

What are we to make of this incredible list of failure to report bad test results, failure to adhere to design specifications, and failure to adhere to protocols? It might be attributable to the internal ethos at Perkin-Elmer. But it seems there were other factors at work as well. Robert Smith in *The Space Telescope* details the myriad political pressures that also were at work. Economic considerations were paramount. Virtually every decision in the early days through the mid-eighties was made to lower costs. Everyone wanted a space telescope, but no one wanted to pay for the real McCoy. At every turn, as we uncover layer after layer, we find that it is the decisions that were made, motivated by varying considerations from political clout to economics to engineering design, that give us the clues to what actually happened. This is sketchy—but the case itself is complicated because the technical nature of the design needs to be part of our conversation, as do the political context and the economic situation. For example, budgetary considerations forced a cutback in the number of NASA inspectors assigned to the project. At every turn we find crucial decisions based on nonuniform considerations that point the way. Finally, we find that neglecting to assess decisions in light of the larger scope of the project is what ultimately led to failure. Why bad tests were not reported and discussed up and down the line is not clear. Why designers were not consulted when production problems emerged is not clear. What is clear is that the engineering design failure was a failure to utilize the feedback function of CPR. Let us be clear about this. The failure was in not following through with the required test to determine if the telescope would work properly prior to launch. This is not the same as saying the telescope was badly designed. It was that the design process was not completely utilized. When the various anomalies reported by Colvin were reported to the appropriate individuals, nothing further happened. What should have taken place was, given the anomaly, to assess the consequences and to factor that into further action. As we have seen, in several key moments this did not happen.

It is important here to pause to consider a possible objection. It might be argued that the launching of the defective Hubble was not an engineering failure so much as a failure of management. Clearly the final decisions not to test and to ignore the results of tests that were already on record were management decisions. But the discussion here is not about engineering versus management. It concerns the design *process*. This was a government contract that had a number of specific protocols built into it. These included consulting with designers when problems occurred, using the null refractor results as intended, etc. What happened was clearly a breakdown in the design process, a process that included both line engineers and management. In short, design is not merely an engineering activity. This is Buchinelli's point. To claim that this was all a management problem ignores the fact that it began with some technicians failing to follow protocol. My concern is how the design process actually works. What we have here is an example of a process that is set out in writing and then ignored by virtually all parties.

In looking at Bucciarelli's and Vincenti's models, with this case in front of us, several things can be noted. First, although I really like Vincenti's structured account of the design process, by leaving out people, he omits the mechanism for moving back and forth between the various stages. Of course people are presupposed—but without specifically involving them, we cannot focus on their decisions, and, as I hope I have shown, it is the decisions that form the *loci* for understanding.

Bucciarelli's move to the social begins to take us in the right direction. But I fear his account suffers by objectifying the object world in a way that delimits human agency. But as should be clear from the HST case, human decisions are at the heart of what happens to the object.

Finally, it is also clear that we cannot argue that these were just bad engineers who were responsible for the problems with the HST. That is, unless you build honesty, apple pie, and the American Way into your definition of a good engineer. These folks knew enough to engage in systematic behavior designed to hide their mistakes and they succeeded, until someone turned on the telescope.

What can we glean from this brief look at this case? If there is a moral, I propose that it is this: the main focus of efforts to understand technological projects needs to be on the people doing the work. It is their decisions that hold the promise of understanding how what happened happened. True, those decisions will be contextualized by a variety of factors, but few of them can be determined ahead of time except in the most general terms—people, materials, institutions, etc.—but that tells us little. The devil is in the details. We need to work at understanding better why people do what they do. In short, we are back to an account of technological explanation as an explanation of the causes and consequences of decisions made by key players.

CHAPTER FIVE

Technology and Ideology

In Chapters 1 and 2 and at the end of Chapter 4 I argued that essential to understanding any innovation or invention are the people involved in its creation and use. One of the reasons for developing this position is the argument that led to the definition of technology as humanity at work. But there is a second reason, equally important. The failure to include the decisions and actions of the appropriate individuals results in philosophical accounts that appear isolated from the remainder of the philosophical conversation, and this is often why philosophers have been seen as having failed to provide an adequate account of technological issues. That is to say, a reason philosophers have been unsuccessful in their treatments of technology is that, for the most part, the questions technologies have raised for them have been addressed in a way that makes it difficult, if not impossible, to integrate them into the broader philosophical discussion concerned with making our way around in the world. There are, in particular, four approaches to analyzing technology from an ostensibly philosophical perspective that have led to the paradoxical result of diverting discussions of technologies away from the broader philosophical dialogue. They are: (1) discussing technology from the context of an ideological bias; (2) attacking or praising some invention simply in terms of whether or not it promotes or threatens some privileged set of moral values; (3) assuming that technology is a monolithic "thing" that is autonomous; and (4) seeing technological innovation as necessarily posing a threat to our political system and our way of life. This last approach may be seen as just a specific instance of (1), but it deserves special treatment since it has an independent history. The remainder of this chapter and the next two are devoted to showing why viewing technologies in these ways is not the proper way to think about technologies if we are interested in exploring their philosophical implications.

Section 1. Some Preliminaries: About Heidegger

One way to contextualize some of the four claims noted above is to see them against the background of the philosophical position of Martin Heidegger. In many respects Heidegger's view embodies these positions, or at least he anticipates them. As we shall see, Heidegger's way of reasoning proceeds at a formally abstract level and it invokes a vocabulary and a way of thinking about technology that make it extremely difficult to understand, no less to integrate, his ideas into a broader discussion. As such he is an ideal candidate for exemplifying some of the complaints we have been working against.

The difficulty of understanding Heidegger is notorious. And, as one might suppose, there are several responses to this issue of difficulty. One can dismiss his writings as simple ravings; one can impute to him various motives, such as deliberate obscurantism or hubris; or one can, as Richard Bernstein does, see Heidegger's idiosyncratic use of language and unorthodox methodology as a deliberate effort to force us to rethink some common notions.[1] This last option is the most charitable way to read Heidegger, and it is the one we will pursue.[2] To do so is to invoke the principle of charity. Basically the idea is this: before you defend or attack an argument or a thesis, do the best possible job of explicating it and making the case for it. For only then will your negative criticisms be considered honest and your positive defense be seen as a serious effort to extend the ideas and not mere sycophantism. Unfortunately, Heidegger has suffered from followers who immerse themselves in his language and ape his philosophical method, only to discredit his views by producing work that sounds profound but that is, at best, Heideggeresque but without much content. Heidegger is not the only philosopher to have garnered his share of disciples who merely mouth the words of the great man and refuse to move beyond his thought. This is a tradition that extends back to Ancient Greece. That it is an old phenomenon is not to say that it is a behavior that we endorse, only to acknowledge sadly that not all that sounds like philosophy is serious philosophical thinking.

Disciples and sycophants aside, the consensus remains that Heidegger is a serious thinker, one of the most important of this century, and therefore

---

1. Bernstein 1993, chap. 4.

2. Though I will break from Bernstein's account by explicating Heidegger without resorting to his special terminology. If a philosophical theory cannot be discussed except by invoking technical jargon, then it is suspect. There is the suggestion that maybe the ideas being tossed about are not germane to anything—that they live in a special world all their own. Remembering that we are trying to draw discussions of the philosophy of technology into the broader philosophical framework, we must avoid jargon and seek a common, understandable language.

it is incumbent upon us to engage him in our discussion, if our view of philosophy as a dialogue is to be something more than empty rhetoric.

The place to start is Heideggger's well-known essay "On the Question of Technology."[3] The structure of the argument is as follows:

1. Traditional accounts of technology see it as a means to an end. This is fine for older technology.

2. This instrumental definition of technology has at its core the concept of causality.

3. Aristotle's fourfold account of technology seems the best way to capture this sense of causality.

4. However, if we are going to use Aristotle's theory of causes, then we should understand it as the Ancient Greeks did.

5. Taken together, the four causes should be seen as being responsible for something else coming to be.

6. Applied to older technologies, we can see technology as responsible for the harnessing of nature, making it available as a set of forces to be used to meet our ends.

7. Modern technology is different from traditional older technology in that the object we are responsible for putting into our arsenal of resources is not just nature, but humanity itself.

8. With the resources of modern science at our command, we are forced (perhaps by the mere having of this knowledge) into turning our knowledge not only against nature but against ourselves.

9. That is why modern technology is a danger.

In Heidegger's own discussion much is made of two distinctions: (a) "true" versus "correct"; and (b) the definition of technology versus the essence of technology. Something can be correct but not true; hence the search for what is true is a function of having a correct account that is

3. When I say that this is the place to start, it should be remembered that Heidegger produced a substantial body of work in what can be called a systematic philosophy. His collected works, the *Gesamtausgabe*, consists of almost 200 volumes. Clearly the position he articulates in "On the Question of Technology" has to be understood in the context of his larger project. An analysis of that project is, however, beyond our scope here.

somehow flawed. This distinction is at the heart of Heidegger's philosophical method. The second distinction frames the structure of "On the Question of Technology." We may have a correct definition of technology as instrumental, but that does not capture the essence of technology. In his search for the essence of technology, he first locates the essence of the instrumental definition in causality. He then asks us to consider what would happen if the instrumental definition were only correct but not true. His search for the essence of modern technology then takes off from reconsidering the original meaning of the Greek account of Aristotle's four causes, leading to the conclusion that because of some uncontrollable power we are forced to treat human beings as mere resources, which is dehumanizing.[4]

While I have difficulties with many of the assertions that come in the twists and turns of the argument, it is the basic assumption, that there is an essence of technology that is uncoverable, that I find fundamentally flawed. It is this same assumption that lies at the heart of the arguments put forth by contemporary social critics. Technology is seen as a thing, a force, in and of itself. It is made into an object against which we can rail. But Heidegger gives no reason nor any argument for the existence of whatever it is that is the force that pushes us into using human beings as resources. By creating an essence for technology, what he has done is to sidestep the real issue, which is: Why do people do what they do to other people? Maybe it is a question he did not want to address or couldn't (see footnote 4). But surely it is no answer to attribute our behavior to some essence of technology.

Further, that the essence of technology Heidegger speaks of is not its definition is in itself an interesting move. Heidegger uses a word to capture what he thinks about technology's power: *Enframing*. But it is just a word, not a definition. And it is a word that really does nothing more that stand in for his whole account. This move makes it impossible really to come to grips with the force of Heidegger's account outside of his own argument. For to deal with it, you have to enter into Heidegger's special way of thinking and his jargonistic use of language, which essentially makes you a captive of his semantics. Once you step outside Heidegger's linguistic web, the force of his way of reasoning loses its punch. For now

4. This is of course the result that Kant argued against—one formulation of the Categorical Imperative being "Do not treat others as means to an end." So it may be that this result argues against a deep-seated Kantian overlay for Heidegger. However, while Heidegger finds the dehumanizing nature of technology disconcerting, his reaction to modern technology is even more puzzling to the modern reader when we discover that Heidegger was a Nazi, a member of the National Socialist Party who enthusiastically endorsed Hitler, the master of dehumanization. See Richard Bernstein's "Heidegger's Silence: *Ethos* and Technology" for an excellent discussion of this dilemma (1993, chap. 4, esp. 118-36).

we can ask: Where did this force that compels us in this way come from? And the answer is that it is an *ad hoc* assumption introduced into the argument through the semantic contortions of the method. By refusing to work toward a definition of modern technology, Heidegger has essentially made the issue unapproachable or, worse still, unchallengeable.

Finally, we need to look at Heidegger's alleged refutation of the instrumental definition of technology. Here is what he has to say in his own words.[5]

> In enframing, the unconcealment propriates in conformity with which the work of modern technology reveals the actual as standing-reserve. This work is therefore neither only a human activity nor a mere means within such an activity. The merely instrumental, merely anthropological definition of technology is therefore in principle untenable. (Heidegger 1954, 327)

It looks as if Heidegger is claiming that technology cannot be conceived of as merely instrumental because it essentially involves nature. But it would seem, by his own reasoning (one hesitates to call it logic), since the result of enframing is to reveal nature as a standing reserve, i.e., to come to understand nature in such a way as to make it possible to harness nature's energies to achieve specific human ends, that the instrumental nature of technology is what is continuously paramount. Further, Heidegger never gives a reason why we must consider the instrumental only with respect to nature—the instrumental can also pertain to the world of ideas, for instance, as in mathematics. So it is not clear that Heidegger has given us anything approaching an argument, much less a conclusive argument, against the instrumental conception of technology.

Despite these concerns, Heidegger's influence is considerable. And it is against that background that we should consider the rest of this chapter.

Section 2. Technology and Ideology

Much philosophical concern about technology can be found in the form of various *value judgments* about the merits of, or threats posed by, particular artifacts or systems. In this sense that form of the philosophy of technology is no more exclusively the work of philosophers than any other set of pronouncements of this kind. This kind of philosophy is nothing

---

5. This is, of course, a translation and not Heidegger in his own German words, which raises an important issue pointed out by Carl Mitcham in correspondence. Heidegger is extremely difficult to translate. And it just may be that the many problems English-speakers have in understanding Heidegger come from the difficulties inherent in translation, exacerbated when dealing with a deeply original thinker.

more than *social criticism*. In Chapter 1, I suggested that the kind of social criticism that characterizes so much philosophy of technology is often difficult to incorporate into the larger philosophical discussion. My reason for making this claim is that the social criticisms I am thinking of rarely make their epistemological and metaphysical suppositions clear. Thus forecasts of dire consequences to follow the introduction of a given innovation generally do not make explicit or even make it easy to figure out what the source of knowledge is upon which the forecast is based or even what counts as knowledge appropriate to these worries. If we don't know what kind of knowledge is being invoked, then it is hard to assess the particular criticism. If we can't assess the criticism, then it is unclear how to incorporate it into our thinking about the way the world works and what actions we should take. This is not an obscure point. Advertisers, for example, know well how important it is to justify their claims about the various attributes of their products. We see this when someone makes a claim that the wonderful effects they are promising for their product are "scientifically proven." When advertisers invoke the aura of science in this way, the audience "knows" that the evidence is reliable because they have already adopted the set of assumptions about the reliability of scientific results. They are then able to assess the claims and their relevance for their own lives, i.e., to decide whether or not to buy the product.

When it comes to understanding the nature of the metaphysical commitments of the social critics, it is often difficult to see where their justification comes from for asserting the causal powers of Technology. Many social critics worry about the existence of something called "Technology" pure and simple, a Heideggerian thing that has its own nonhuman source of power and direction. It is supposed to control our lives. It is not identical to the specific innovations and inventions whose genesis and introduction into society we can trace and examine. But I find the social critics' perspective puzzling. What does it mean to say, for example: "Technology is taking over our lives?" One way to understand this claim is to interpret "Technology" as a thing in itself that has its own set of causal powers and operates on its own, independent of human interference.

Why is it that these purportedly philosophical investigations into technology are, in a certain sense, unreachable by the general philosophical community? One answer is that most of this form of social criticism is framed in the context of a specific ideology that presupposes the intrinsic merits of a particular value or set of values, usually moral values, and that frames its perception of the world in terms of how these values are being thwarted.

Ideologies are interestingly negative. People who employ them are generally concerned with imputing motives to other people or things they believe are keeping them from Utopia, where the full-blown consequences of their particular utopian visions are rarely explored or exposed by their advocates. Consider the following example, paying particular attention to the rhetorical language. The passages are taken from Langdon Winner's *The Whale and the Reactor*. My objections to Winner's claims are two: (1) many of his arguments are flawed and unsupported; but most important, (2) his discussion is unrelentingly ideological. That is, he insists on characterizing specific complex systems of artifacts as the embodiment of ideologies. I find this a strange claim. Against it, I argue that tools and technical systems *are inherently ideologically neutral*. Individuals with particular axes to grind may employ a tool to achieve their ends, but this does not make the tool itself ideological.

In the selection below, Winner is relaying his reactions to a visit to the site of the Diablo Canyon nuclear reactor on the coast of California, which just happens to be near his boyhood home:

> Feelings of anticipation swelled in me as the bus rolled to the top of the last hill separating us from our view of the Diablo Canyon site. As it reached the summit of a small plateau, I looked out over a vista that sent me reeling. Below us, nestled on the shores of a tiny cove, was the gigantic nuclear reactor, still under construction, a huge brown rectangular block and two white domes. In tandem the domes looked slightly obscene, like breasts protruding from some oversized goddess who had been carefully buried in the sand by the scurrying bulldozers. A string of electric cables suspended from high-energy towers ran downhill, awaiting their eventual connection to the power plant. In the waters just off shore lay two large rocks surrounded by a blanket of surf, as elegant in appearance as any that one finds along the Pacific coast. One of them, Diablo or "Devil's" Rock, loomed as a jagged pinnacle. Next to it, not too far away, was a smaller, but even more finely sculptured piece of stone, Lion Rock, which looked very much like a lion at rest or, more accurately, like the Sphinx itself reclining on its haunches, paws outstretched on the surface of the ocean, silently asking its eternal question. At precisely that moment another sight caught my eye. On a line with the reactor and Diablo Rock but much farther out to sea, a California grey whale suddenly swam to the surface, shot a tall stream of vapor from its blow hole into the air, and then disappeared beneath the waves. An overpowering silence descended over me.
>
> * * * * * *
>
> Although I had known some of the details of the planning and construction of the Diablo Canyon Reactor, I was truly shocked to see it actually sitting near the beach that sunny day in

> December. As the grey whale surfaced, it seemed for all the world to be asking, Where have you been? The answer was, of course, that I'd been in far-away places studying the moral and political dilemmas that modern technology involves, never imagining that one of the most pathetic examples was right in my hometown. My experience with the reactor itself, seeing it at a particular time and place, said infinitely more than all of the analyses and findings of all the detailed studies I had been reading ever could.
>
> About that power plant, of course, the standard criticisms of nuclear power certainly hold. It does pose the dangers of catastrophic nuclear accident similar to the one at Three Mile Island. Certainly it will produce routine releases of low-level radiation and thermal pollution of the surrounding ocean water. No one has developed a coherent plan for storing long-lived radioactive wastes that this plant and others like it will generate. Already ten times more costly than originally estimated, its $5.5 billion price tag does not include the tens of millions of dollars it will cost to decommission the thing when its working life has ended. From the point of view of civil liberties and political freedom, Diablo Canyon is a prime example of an inherently political technology. Its workings require authoritarian management and extremely tight security. It is one of those structures, increasingly common in modern society, whose hazards and vulnerability require them to be well policed. What that means, of course, is that insofar as we have to live with nuclear power, we ourselves become increasingly well policed. (Winner 1986, 165, 175)

It is perfectly clear that Winner is advocating a definite point of view. There is nothing wrong with that. Winner is opposed to nuclear energy. But we have to wait until he lists the "'standard' criticisms of nuclear power" before we get either an argument or any factual information that might possibly back up his point of view. And even at that point, the manner in which he advances his reasons should cause the innately cautious reader to become suspicious. The cause for suspicion is quite straightforward: Winner lists the criticisms as if they were established facts. But just because someone offers up a criticism of something, it doesn't follow that the criticism is valid. Let us take a look at his "standard criticisms."

Yes, it is certainly true that there is a danger of "catastrophic nuclear accident." But what counts as "catastrophic"? And more important, how high is the probability of the catastrophe occurring? Consider Winner's own example, Three Mile Island (TMI). Regarding his claim that the reactor at Diablo "pose(s) the danger(s) of catastrophic nuclear accident similar to the one at Three Mile Island," how many TMIs have there been?

One accident shows that such occurrences can happen in those circumstances in which they did happen. It does *not* show that there is an equal probability that the same accident will happen in any similar such set of circumstances. Nor does it allow for the possibility of people learning from their mistakes, thereby correcting for and eliminating the problems that caused the first situation.

Second, what does he mean by "catastrophic"? TMI did not result in any provable loss of human life. The accident at Chernobyl did. (Granted, Chernobyl occurred after Winner's book was written.) TMI *could* have been a catastrophic disaster, but compared to Chernobyl it was merely an accident. The problem here is that if you are already opposed to nuclear energy, then it is not difficult to see TMI as the embodiment of all your worst fears. But calling an accident a catastrophe doesn't make it one.

Third, it is misleading to assert that "certainly it will produce routine releases of low-level radiation." It is misleading because Winner provides no evidence to support this claim and asserts it as if it were incontrovertible, which it is not. The manner in which this claim is presented suggests that every nuclear plant has been tested and that it can be shown that this in fact happened. First, this is not true. Second, it ignores the existence of low levels of normal radiation in the general environment, suggesting that low-level radiation represents a novel situation, which it does not. Likewise, it is false to assert that no coherent plan has been developed for storing waste. There is at least the plan to use deep storage in salt caves. Winner may not like the plan, and he may have his own reasons for objecting to it, but the flat assertion that no such plan exists is simply false.

If we turn away from the false and/or misleading factual claims and turn to the rhetoric, the ideological stance becomes more obvious. It is clear that this is not an unbiased account, but, of course, it is not intended as such. The language is designed to stir certain feelings in the reader and elicit specific reactions. Consider again the first selection quoted above. This time I have italicized some of the more suggestive phrases.

> *Feelings of anticipation swelled* in me as the bus rolled to the top of the last hill separating us from our view of the Diablo Canyon site. As it reached the summit of a small plateau, I looked out over a *vista that sent me reeling*. Below us, *nestled on the shores of a tiny cove*, was the *gigantic* nuclear reactor, still under construction, a huge brown rectangular block and two white domes. In tandem the domes looked slightly *obscene, like breasts protruding from some oversized goddess* who had been carefully buried in the sand by the *scurrying* bulldozers. A string of electric cables suspended from high-energy towers ran downhill, awaiting their eventual connection to the power plant. In the waters just off

> shore lay two large rocks surrounded by a *blanket of surf, as elegant in appearance* as any that one can find along the Pacific coast. One of them, Diablo or "Devil's" Rock, loomed as a jagged pinnacle. Next to it, not too far away, was a smaller, but even *more finely sculptured* piece of stone, Lion Rock, which looked very much like a lion at rest or, more accurately, like the Sphinx itself reclining on its haunches, *paws outstretched on the surface of the ocean, silently asking its eternal question.* At precisely that moment another sight caught my eye. *On a line with the reactor and Diablo Rock but much farther out to sea, a California grey whale suddenly swam to the surface, shot a tall stream of vapor from its blow hole into the air, and then disappeared beneath the waves. An overpowering silence descended over me.*

There are two different sorts of things going on in this passage. First, there is the pitch to the emotions. To see this, we need only focus on such phrases as "anticipation swelled in me" and "vista sent me reeling" or the final two sentences. The second thing Winner is attempting to do in this passage is to invoke an image of the contrast between the serenity and dignity of nature and the obscenity and alien nature of the human artifact, the reactor and the power lines. Unfortunately, he ruins his own effect by invoking the image of the goddess and coupling it with the obscenity of the reactor. The goddess image is one that has been nicely tied to the mythology of nature and the nourishing image of the mother.[6]

In the first instance, where Winner's choice of language is clearly rhetorical, it is safe to say that he is making a series of explicit value judgments. In the second case, where he is trying to contrast man and nature, with the objective of stressing the offensive aspects of human action, he is pushing an ideology. His ideological stance can be seen again at the end of the long section quoted earlier, when he talks about how the existence of nuclear plants leads to our being well policed.

Let us consider the questions of ideology and values separately. First I look at some of the problems ideological positions present, then I consider the role of values in Section 3.

With regard to attacking or defending technology from an ideological basis, the problem is that we can never make philosophical headway by taking an ideological stance. To do so precludes our ability to resolve whatever disagreement may be at issue, as well as our ability to understand the artifact or system involved.

---

6. C.f. Camille Paglia, *Sexual Personae* (1990). In this case either Winner has made a mistake by identifying the reactor with the goddess, or perhaps his subconscious is telling us something other than what he appears to be trying to say, namely that nuclear energy is not so unnatural.

The notion of an ideology is somewhat slippery. To characterize an idea or a social phenomenon as an expression of an ideology of some kind is already to take an adversarial role with respect to that phenomenon. More precisely, it is to assess that phenomenon as a product of a value system to which you are opposed. In other words, if your opponent is a capitalist, you see his complaints as nothing more than so much ideological hogwash, while your point of view is, of course, merely good common sense. In the broad sweep of ideas this can be expressed by saying that an ideology is essentially a conceptual scheme for interpreting and making sense of events. This is sort of like wearing a special pair of glasses that are sensitive to infrared and allow you to see the world in a way you would not otherwise. The major difference between ideological glasses and a conceptual scheme is their uses. In a certain sense we *require* a conceptual scheme to comprehend the world *at all* and to act in it. Perhaps "comprehend" is too strong, but we at least need such a framework to make sense of what we see and do. For example, we have to be taught that this is a chair and that is a cow and not to hit your baby sister.

What counts as a chair is not obvious. Chairs do not scream out at you "I am a Chair!" The items we identify as chairs are so considered because over a period of time our culture has evolved a set of criteria that provide us with the conceptual equipment and language to identify this as a chair and that as a cow. The criteria are not always obvious; we learn them at our mothers' and fathers' knees as we learn the language, and they find their way into the language by an informal process of interaction and negotiation among various players arguing about the world and what is in it. When we identify an object as a chair, we are employing the results of that process as well as demonstrating that we have learned this language and know how to use it properly—meaning by that that we have mastered the rules of the language and its various criteria for identifying objects. To have accomplished this is to have command of a conceptual scheme.

To call an ideology a conceptual scheme does not by itself take us very far. There are different kinds of conceptual schemes and not all are adequately characterized as ideologies unless, that is, a particular conceptual scheme, seen as a specific interpretation of some feature or features of the world, is itself being attacked by an ideology. For example, there is a sense in which the theories, facts, and explanations of science can be described as the results of a particular kind of conceptual scheme, namely one in which the major grouping of concepts—call them categories—are designed to (or perhaps simply evolved so as to) capture fundamental features of nature like space, time, causation, etc. When using the conceptual scheme of science, we employ these categories to structure perceptions and understandings of the natural world.

Science, described as such a conceptual scheme, is not thereby an ideology. But it *can* be so *described*. (Of course, the fact that it can be described as an ideology does not make it one. Likewise for all descriptions, for the fact that we describe the world in a certain way does not mean that the world is necessarily that way.) Consider an attack on the conceptual scheme of science based on feminist thought, where the claim is that the categories of science represent a specifically *male* perspective on how to carve the world at its joints.[7] If the case can be made that science does indeed embody such a male perspective, then the feminist would be well on his or her way to successfully characterizing science as an ideology. Why is this so?

It would seem that the ideological character of science would derive from the (alleged) fact that since its categories represent an exclusively male point of view, it is not a universal perspective, as some claim it to be. The fact that it claims to be universal and may not be so is not the crucial point here. What is crucial is that if science cannot represent a universal (perhaps "objective") point of view, then the categories that form the basic constituents of its framework must not be obvious or correspond to the ways things really are and, hence, must be the results of decisions based on certain contaminated assumptions. The heart of these assumptions is the adoption of a certain perspective, just as the heart of Marxism is historical materialism and the heart of a modest feminism is the vision of total equality for women and men.

The adoption of a certain perspective by itself is not enough to differentiate conceptual schemes from ideologies. And it just may be the case that in terms of their semantic structure they cannot be differentiated. Nevertheless, we can still distinguish between them in terms of their *use*. In particular, it is the quasi-pathological use of a conceptual scheme that turns it into an ideology. Thus, despite the crucial importance of its basic assumption that women and men ought to be treated equally in all matters, feminism is misused and indeed perverted when its defenders find the cause of each and every evil that befalls women to be the result of male concerns with their subjugation. Likewise, science is misused when its methods and discoveries are declared to be the final arbiter of truth, no matter what. And when this happens, whatever intellectual authority that particular perspective on the world initially may have had rightly dissipates.

There is another equally important aspect of ideologies that captures another facet of the quasi-pathological misuse of conceptual schemes. (And here it is important to remember that it isn't the conceptual scheme that is being faulted, but rather the people who are misusing it.) That is the ability of an adherent of the ideology to explain all relevant events

---

7. For a small sample of the different female feminist criticisms of science, see Haraway 1991 (esp. chaps. 1-5); Harding 1986; Longino 1990.

exclusively in terms of the basic assumptions of the ideology. Thus, from a creationist perspective, all natural events are to be understood in the light of their particular interpretation of the Bible. The mark of a successful ideology is that not only can it explain all relevant events in its own terms, it can also resist and possibly even be seen to refute alternative explanations based on other perspectives. This is done by either rejecting the formulation of the attack or reinterpreting the issue in its own terms. In short, the mark of a quasi-pathological ideology is that its claims cannot be shown to be false.

So, if such ideologically based claims are not falsifiable, how do you resolve a debate between adherents of two different ideologies? You can't. And if our concern is to deal with technology from a philosophical perspective, one that permits debate and allows for alternative analyses, then the moral is that talk of technology couched in ideological terms renders philosophical discussion impossible.

"But," you may argue, "that is very fine and dandy for you to say, yet how are we to avoid taking an ideological stance if, as you said earlier, conceptual schemes are essential for thinking about the world?" Good question. One way to answer it is to ask another: How does an otherwise innocent conceptual scheme turn into an ideology? We already know part of the answer: by the way in which it is used. But that just leads to another question: What is it that makes it possible for conceptual frameworks to be used or misused in this way? The answer here is basically simple, but it takes a bit of explaining.

In order to make our way around in the world, we use a *number* of conceptual schemes. Together they constitute a conceptual *framework*. Each conceptual scheme is domain-specific, that is, its concepts and categories are used to reason about a specific domain of objects or concerns. For example, the scientific scheme is concerned with the natural world; our aesthetic scheme gives us the resources with which to discuss beauty, proportion, and style; our moral scheme establishes the mechanisms we use to create and evaluate social relations. Unlike the conceptual schemes, which are all domain-specific, our conceptual *framework*, being composed of the sum total of our conceptual schemes, has no one specific domain. Conceptual frameworks evolve over time and they form a close interactive relationship with language. We reason in a language, and the vocabulary of that language provides the names for the concepts and the grammar for formulating our thoughts in that framework. As such, conceptual frameworks are the products of social interactions and they are transmitted from one person to another through the teaching and learning of language, which entails both knowing how to use a language and how to criticize its misuse. By retaining and using terms in specific ways, conceptual frameworks also embody a certain kind of knowledge about the world generated by the group in whose language it is formulated. For that group, in that language, that knowledge is seen as common sense. The development of any specific conceptual scheme will, therefore, occur in the context of the principles of common sense. This is not to say that the result will be commonsensical; indeed we are currently exploring how it is that from a commonsense basis a conceptual scheme gets misused. Hence science is an outgrowth of common sense, a refined approach to empirical experience of the world (Pitt 1981). But eugenics, as a specific type of science, may be a misuse of the principles and goals of science as a whole.

Once we see how conceptual schemes function to deal with specific domains of concerns within the larger conceptual framework, we can then see how the concerns of one scheme can sometimes form the basis on which to use another scheme. First, from the fact that each conceptual scheme has its own domain of concern, it doesn't follow that that domain is its exclusive concern. Thus the relations among objects in the world may be the legitimate concern of science, but they also can be the concern of art. Second, the results of the applications of a conceptual scheme may themselves become the object of another scheme. This is what happens when we criticize the results of science from a moral point of view.

If this account is correct, then I propose that a conceptual scheme becomes an ideology when it becomes subordinate to a *particular moral* scheme, for the key item in an ideology is the normative force it brings with its perspective. Ideologies are ways of asserting what the *proper* relations between and among people and the world *ought to be.* In this sense they are fundamentally moral schemes, but in their intermediate form they take the shape of political agendas. By that I mean nothing more than that an ideology expresses a certain perspective on the power relations among people from a given moral point of view; in particular, one that will not tolerate any competitors.

The answer, then, to our question of how to avoid using ideologies as the basis for understanding the philosophical issues surrounding analyses of technology is: *avoid casting all concerns about technology as political issues.* And the justification for this claim is that there is no general principle or rule in our conceptual framework that gives us the license to subordinate any conceptual scheme to any other.[8] This is not to deny that conceptual schemes are in fact appropriated for various uses; it is merely to say there is no obvious philosophical justification for such appropriations.

---

8. For an elaboration of this view of the relations between conceptual schemes, see Pitt 1981, chap. 5.

"But," it might be argued, "while your advice to avoid approaching the discussion of technology from an ideological stance seems reasonable, since we don't want to find ourselves in a position where we can't agree on what to do, might it not be the case that the technology itself is an expression of an ideology. For example, isn't Langdon Winner correct when he points out that power plants are constructed by individuals deliberately concerned to disenfranchise the masses or ordinary people by forcing them into a position of dependence on the power company? Hence, are we not forced to deal with this technology in equally ideological terms if we are to control it? After all, the political domain is concerned with the exercise of power, and the control of technology involves just that."

If we are to avoid being ideological about ideology, surely we can't dismiss this objection out of hand. Nevertheless, it is not at all clear what it means to say that technology is ideological. If, for example, we consider Winner's nuclear reactor as "an inherently political ideology" (Winner 1986, 175), what does that say about the technology itself?

Consider what's at issue here. If we want to know the respect in which a technology is inherently ideological, we could be asking one of two questions, or possibly both. (1) What are the intentions of the people who designed and brought the reactor on line? For example, are they concerned with providing a necessary service, or are they concerned with controlling people? (2) What is the best interpretation, *after the fact*, of the consequences of implementing such a complex system? Both of these questions are important. But they are fundamentally different. In the first, the assumption is that the system is ideological if we can show that it was constructed for the *ideological* purpose of controlling people in specified ways. In the second, given the system and given *our ideological concerns*, how can we best explain the impact of that system on our lives in terms of our own ideological ends? In the case of the first question, we might be able to claim that the power plant is inherently ideological if we can show that the people who planned it and paid for it really had evil intentions such as Winner suggests. For the second, we need to show how *the effect* of having this system on line is only to keep the masses subjugated. These questions are important not because they give us a clue as to whether or not a given innovation of system of tools and artifacts is the expression of an ideology, but because these questions point us toward something more fundamental, namely, the direction of important philosophical questions about technology. These questions concern, first, the procedures and factors that were at work in the decisions and the options that led to the development and use of, e.g., the Diablo Canyon reactor. In addition, there is the question of assessing the consequences of using nuclear reactors to generate electricity.

It would appear that, at best, ideology enters at the point of assessment. For it seems almost unavoidable to assess the effects of a given set of tools in ideological terms. However, and here's the rub, for any given ideologically based assessment, it is possible to provide an alternative in terms of a different ideology. Thus, on Winner's account, the use of nuclear reactors "requires authoritarian management" and, consequently, our being policed. But, I would argue, isn't the alternative equally unacceptable? No plant, no electricity; no electricity, there go the benefits of modern society. But the benefits of modern society are what make human growth possible. Thus to be free from the daily drudgery of having to provide basic services for oneself and one's family is what allows for science, art, music, and even philosophy. Arguing in this fashion invokes yet another ideology, one that extols the benefits of technology and its liberating consequences. More to the point, it is an ideology based on certain values, e.g., that reflective thought and its expression are very important to my way of life.

But irrespective of which ideology is being employed, the important point for our analysis is this: despite the fact that Winner posed the questions in his unique polemical manner, he got the key issues right. That is, the interesting questions about such complex systems are the questions about how they came to be constructed in the manner in which they did and how best to assess their merits.

If we are concerned with assessing the consequences of employing a complex system such as the electric power generating system, then an analysis of the intentions of the individuals involved in developing and employing that technology is beside the point, since those worries should be part of our analysis of the decision structure used to design and implement the system. There are occasions when it would appear reasonable to address the question of whether or not the system did what it was designed to do. But while that kind of question is reasonable, it does not entail that the *only* way to answer it is in ideological terms. So, while it is possible that one aspect of assessment might involve uncovering an ideological component to the original intentions of the designers and decision-makers behind the technology, it does not follow that assessment must proceed in those terms alone. In the case of Langdon Winner's reactor, I doubt that the motive of those who built it was to control people's lives. I suspect financial gain and a demonstrated demand for electricity were at issue.

On the other hand, one objective of an assessment might be to determine whether or not the ideological aims of the decision-makers (assuming there are such) who developed the system were met. Thus we might try to determine if the implementation of certain technological devices and systems have imposed ideologically driven constraints on the individuals using those devices and systems. That is, by using electricity generated by nuclear reactors owned by large electric companies, are we being ideologically constrained in an ideologically desirable way? That is an important question in assessing the merits of the system from the perspective of its designers, should they have ideological axes to grind. But even here it is important to note that it is not the system that is being assessed for its ideological purity but, rather, *the effects* of the system. Furthermore, it seems fairly obvious that, no matter what the device or system is, if it is put to different uses with different ends in mind, the consequences of using it will be different. This suggests that even if a tool is used by individuals committed to a particular ideological stance, it is not the tool that is ideological. I conclude, therefore, that technology, tools, and systems of tools are ideologically neutral.

Section 3. Technology and Values

From our analysis in Section 1, it is increasingly clear that important philosophical questions about technology are concerned with (a) the *decisions* people make when developing and implementing innovations, and (b) *assessing the consequences* of using tools and social systems. Both of these issues involve making *judgments*. If we turn to the decision processes that lead to the establishment or implementation of certain innovations, there are two separate kinds of issues to address. The first concerns understanding the decision-making processes themselves. In examining this issue, we need to distinguish values from ideologies.

The second problem concerns the empirical question of what actually is going on, i.e., who is deciding what to do and why? This, however, does not reduce to merely an empirical question of identifying the specific individuals involved. It is a philosophical question, at least the "why" part. The heart of the matter is: Where do values enter into deliberations that eventuate in new tools and ways of doing things? Deliberations eventuate in judgments—judgments assert values. They enter at any point where a decision is being made. They enter at the point, for example, where we decide to use one decision-making process to resolve a question rather than another. They also enter at the point where we decide to allocate resources to resolve a problem by developing some technique and when the person charged with devising the technique decides to use one hypothesis as a basis for his project rather than another. This point is an old one and it was driven home by Richard Rudner in his analysis of the role of values in scientific decision making (Rudner 1953). Judgments are *ipso facto* value judgments. The making of a judgment is a determination of the importance, priority, or relative merit of a given point of view, all or any of which invoke values.

But to invoke values is not necessarily to invoke ideology. Rather, to invoke values is to appeal to priorities, goals, and objectives. Part of the problem behind assuming that all value questions are ideological derives from an equally generally unstated assumption—namely, that all values are moral values. But that is simply not so. In addition to moral values, there are aesthetic values and cognitive values, just to name two other kinds. The role of cognitive values is often overlooked when technology is viewed only in ideological terms. Thus when we assert that technology is ideologically laden, we imply that the only values that enter in or that are relevant to the assessing of discreet innovations and systems of such or to their implementation are values that are associated with some ideology, hence, with political philosophy, hence, in some derivative fashion, with morals. This need not be the case.

There is a domain of inquiry equally laden with values that is not ultimately moral but rather concerned with the *cognitive* dimensions of human life, that is, with those aspects of our lives that are involved in the generation, assimilation, and use of *knowledge*. Thus the decision to employ a certain system could be made on the basis of cognitive concerns rather than values associated with the moral domain, e.g., our decisions to build space-faring probes and to explore the solar system because we have "knowledge" as a cognitive value. By overlooking or ignoring cognitive values, we bypass the *epistemological* questions associated with inquiries into technology. And, I suggest, in so doing, we miss the opportunity to bring these inquiries into the broader philosophical discussion.

In sum, there seem to be two basic problems associated with approaching the philosophy of technology from an ideological stance. First, there is the false assumption that the value-laden dimensions of areas of discourse associated with technology are overshadowed by ideological considerations and can have no other dimensions beyond those.

Second, approaching the analysis of technology ideologically eliminates the possibility that claims of ideological import can be tested empirically. If this is the case, it may be all well and good and even interesting to some, but the important question now is: Of what consequence? It is important for the following reasons. We need to know a number of things about our tools and their delivery systems. We need to know how it came to be that a given innovation arose. What were the factors, political, scientific, engineering, economic, that resulted in the building of the Space Shuttle, for example? Second, we need to know the effect of implementing a given system. How, for example, does the increasing bureaucratization of government affect our freedom or our individuality? There are many kinds of questions we need answered, some of which can be answered by empirical research. Others demand further philosophical analysis and clarification. The value of de-ideologizing the philosophy of technology is that it allows the empirical questions to be sorted out from the questions of values.

A consequence of taking this approach is that it strips away the romanticism that we have built into some of our discussions concerning technology. For example, we can no longer talk about "technology taking over our lives," since our attention has been directed to the decisions men and women have made with respect to developing, building, etc., a satellite system. It forces us once again to shoulder the burden of responsibility for the decisions that result in the construction of systems and the construction and marketing of new products, and for assessing the impact of the results of such developments. It leads away from talk of impersonal and indeterminable forces governing the evolution of technology, and once again asks the question: "To what extent do the accidents that are the normal produce of human commerce affect our attempts to increase the quality of life?"

I return to a theme I have been elaborating here: if we are to understand the world, we must begin with the assumption that it is people who are involved in the creation and the alteration of the environment. To be sure, there are man-made disasters. There are also natural disasters. But to the extent that human beings respond to the environment, be it the natural environment or the environment they created, and to the extent that human beings attempt to change or control or alter that environment by the building of dams or the implementing of governmental regulations, then we must start with the decisions individuals make, and not with a mysticism that clouds our ability to improve upon those techniques that we have decided are not to our advantage.

Turning to the question of the tool or a system of tools and techniques itself, we have already begun to explore the view that any particular tool or system is neutral with respect to ideological concerns. Unfortunately, this claim is ambiguous. It might be interpreted to mean that the individuals involved in designing the tool or system have no ideological axes to grind. But surely this is false. It seems perfectly reasonable to assume that some tools are designed with specific ideological concerns in mind. By way of example I would suggest that the First Amendment to the U.S. Constitution might be viewed as securing the development of a free press because of a commitment to the idea that such an institution is the best safeguard of democracy. This also could be an example of the most uninformed kind of assessment, but I submit that it is of the same kind as most ideological claims. More damaging is the acknowledgment that the concerns of those responsible for the development of any given tool are beyond our ability to recreate with any degree of certitude. That is, can we ever be sure of the intentions of the authors of the First Amendment, and furthermore, is our knowledge of their intentions really relevant to our current concerns?

A second interpretation of the idea that any given tool or system is ideologically neutral is the following somewhat more complex view. Whatever the ideologically motivated intentions of the originators of an innovative system, the mechanisms of both the decision-making process and then the system itself have the effect of neutralizing those initial presumptions. Consider the situation in which *what* an individual wants to accomplish can't be done under the given circumstances, but "if we give a little here and push a little there, then we can get *most* of what you want." This particular scene can be filled out in many different ways—products can't be produced according to the designer's vision because of cost, but if we just adjust here a bit, . . . etc.

The final point is the case in which the idealized perception of the system has a bearing on the changing attitudes of the participants in the decision process. One place to see how this works is in the Supreme Court of the United States. Lest there be objection to using the courts in the context of technology, I remind the reader that the judicial system is a tool for adjudicating social conflict, a technology if ever there was one!

In the case of understanding judicial decisions, we cannot underestimate the extent to which the values of the judges themselves are affected by their participation in the decision-making process (Schubert 1965). The case of our courts of law is a good one, for it points out a number of features. The Supreme Court has no enforcement powers. The acceptability of a decision by lower courts is a matter of individual judges deciding to abide by the decision of the Supreme Court. So to begin with, a decision of the Supreme Court will be tempered by the realization that the decision must be couched in acceptable terms. Second, assuming that the judges have individual ideological agendas to enact, having the Court produce a pure ideological manifesto will be impossible if only because it will be the product of the forge of compromise among the members of the Court. In addition, individual judges may discover that, in the process of hammering out a stand on a variety of other issues, previously privileged views on some other matters may have to be abandoned. That is, the realities that confront the judges on specific cases may force them to consider as impractical other beliefs they may have about the way the system ought to function.

Now if, under circumstances such as these, it is claimed that the decision-making process is inherently ideological—so be it. In this system, values, seen as an expression of unactualized but highly desired ideal states of affairs, are confronted by the realities of the world around us. It is in such confrontations that both the values and the world change, pushing and shoving on one another, testing for weak spots. But if in the process of attempting to implement certain values, the values are themselves changed, then it is hard to determine where the ideological bent is to be located. If the process defuses the ideological bomb by the way it forces compromise and change, then the process itself cannot be seen as ideological—hence, we would do well to consider it neutral.

CHAPTER SIX

The Autonomy of Technology

It might seem that it is but one step from the view that technology is ideologically neutral to the view that technology is autonomous. If, as we noted in Chapter 5, a tool or system can contribute to the decision-making process by forcing changes in values, then surely, it might be suggested, the system itself becomes an independent actor in the process. Maybe so, but probably not. But the view that technology is autonomous is a popular one. Consider what Jacques Ellul has to say on the subject:

> –Technique is autonomous with respect to economics and politics
> –Technique elicits and conditions social, political and economic change. It is the prime mover of all the rest, in spite of any appearance to the contrary and in spite of human pride, which pretends that man's philosophical theories are still determining influences and man's political regimes are decisive factors in technical evolution. (Ellul 1964, 133)

Ellul may be right about the role philosophical theories and political regimes play in technical evolution, but his claims also sound somewhat exaggerated. More important, the kind of claim he makes for the autonomy of technology makes it sound as if it were unfalsifiable, especially given assertions such as "in spite of any appearance to the contrary."

Unfortunately, claims like Ellul's have become commonplace. They amount to treating technology as a kind of "thing," and in so doing they reify it, attributing causal powers to it and endowing it with a mind and intentions of its own. In addition to the fact that it is empirically false that *Technology* has these characteristics, reifying Technology moves the discussion, and hence any hope of philosophical progress, down blind alleys. The profit in treating Technology in this way, to the extent there is any, is only negative. It lies in removing the responsibility from human shoulders for the way in which we make our way around in the world. Now we can blame all the terrible things that happen to us on Technology! It is only after the first moves have been made toward reifying Technology that we hear about such things as the "threat" of technology taking over our lives. Likewise, reification leads to misleading talk about technology being the handmaiden to science, or some variant on that theme. In other words, reification makes talk about autonomy possible. But, I will argue, it is a major mistake to think there is any *useful* sense in which we could conceive of technology as autonomous.[1]

It is important to stress the "useful" here. It is no doubt possible to contrive outrageous examples to show there is something called autonomous technology. But before we allow misdirected philosophical analysis to take us into the world of science fiction, we can at least take the time to understand what is really going on. Technology, even understood in its more popular-culture sense as new gadgets and electronics, among other things, is such an integral part of our society and culture that unless we ferret out the ways in which these devices are actually embedded in our lives, we may fall victim to a kind of intellectual hysteria that makes successful dealings with the real world impossible. The first step to take if we are to avoid this danger is to clarify the kinds of issues that can reasonably be addressed. To a large degree this means separating the significant from the trivial.

Section 1. Trivial Autonomy

There are at least two cases of talk about the autonomy of technology that are non-starters. That is, if these popular topics of discussion are considered carefully, they easily can be shown to be irrelevant to serious consideration of the issue, since the kind of autonomy they address is trivial.

In the first case, some version of the following account is given of what it means for technology to be autonomous: technology is autonomous when the inventor of a technology, once the technology is made available, loses control over his or her invention. The development of the digital computer can be used as such an example. Once computers entered the public domain, it was impossible for anyone to call them back. The rapid increase in their sophistication and the all-pervasiveness of their employment in society made it impossible to avoid them once they entered the marketplace. Surely, the story goes, this is a case of autonomous technology.

Well, yes and no. Yes, it is autonomous, if by that is meant only that the inventor alone can no longer control the development of the technology.

1. For examples of this "style" of philosophizing about technology, see all of Ellul 1964; Winner 1977.

But this is a trivial sense of "autonomy," since it is true of all aspects of our society. Once in the public domain, each item is beyond the control of its inventor in some sense or other. But that does not make the item autonomous. Its further development is a direct function of how people employ it and extend it. To the extent that people are necessarily involved in that process, the invention cannot be autonomous. Rather than being conceived of as an independent agent that acts on its own, the invention is seized opportunistically as a means to an end. It is used, changed, augmented, or discarded, depending on the goals of the agents. That these various uses were not envisioned or intended by its inventor does not make the invention autonomous in any interesting sense.

The second trivial case of autonomous technology concerns the consequences of innovation. Here it might be claimed, for instance, that because the inventor of a device or system failed to see the consequences of employing it in a certain way, the item has a life of its own and is autonomous. Thus it would appear on this scenario that the use of nuclear plants to generate electricity is evidence for the autonomy of nuclear energy, since this use was not foreseen by Einstein in his famous letter to President Roosevelt informing him of the wartime potential for nuclear energy. This, too, is an incorrect conclusion. The fact of the matter is that no one can foresee all the consequences of any act. That fact, however, does not entail that once some action is taken, the consequences of that action are autonomous. That the full consequences of introducing large-scale manufacturing techniques for the production of automobiles were not anticipated by Henry Ford does not mean that those consequences were due to the automobile or to the processes, economic, social, and engineering, that produced it.

The key to understanding this second point lies in realizing that once an invention or innovation leaves the hands of its inventor, it also leaves behind the circumstances in which the actions of only one person can affect its development and employment. Once it enters the public domain, its diffusion generally will be the result of community decisions; and as we noted, these are the kinds of decisions that are the results of compromises. That there is no logical order to the patterns these decisions take should come as no surprise. Compromise is a function of a variety of factors, and it is impossible to tell in advance which of them will be persuasive in any given situation. Furthermore, it may be that *it is this lack of absolute predictability with respect to the outcome of community decisions that itself produces the illusion of the autonomy of technology*. But the fact that the role an innovation acquires in a society is a function of complicated community decisions, which decisions are at best compromises (at worst they are the results of collusion and corruption, which themselves involve compromise), does not entail that the innovation is autonomous. *Quite the contrary*. Given the kind of buffeting and manipulation this process involves, it would appear that it would be anything but autonomous!

Thus arguments from the eventual lack of control of the inventor and the failure to foresee all the consequences fail to secure the case for the "autonomy of technology." But there are also other arguments we need to consider.

Section 2. The Process of Technology

Well-intentioned writers and critics have commented on various aspects of technology which they see as raising the possibility of a serious sense of autonomous technology and, along with it, the specter of apocalypse. One of the best examples of the kind of worry expressed by these authors can be found in John McDermott's essay review of Emmanuel Mesthene's *Technological Change*, "Technology: The Opiate of the Intellectuals" (McDermott 1969). In that review McDermott speaks of a kind of momentum certain devices or systems acquire, thereby providing the appearance of autonomy.

Consider the following McDermottian scenario. A growing retail company located in Fairbanks has just hired a fancy up-to-date accountant with an MBA to manage the financial records of the company, which records are currently in a condition closely resembling chaos. Our accountant is a bright young urban professional. Given the size of the company and its projected growth, she argues persuasively that in the long run it will be cheaper and more efficient to buy a couple of computers than to hire additional staff and to continue handling the books in the traditional way, with ledgers entered by hand, etc. She produces a report showing the projected costs of people versus machines, calculating only for the long run the cost of benefits and retirement for the people and maintenance for the machines. She wins her case and the computers are purchased. But once the computers are introduced, air conditioning is not far behind, because the computers need a cool environment to function optimally. But, our fictional tale continues, air conditioning simply can't be added on to the current structure housing the company offices. Either we redesign the old building to handle air flow and pressure, or we look for a new one. Finally, our storyteller says with a knowing look, the president of the company is totally confused and dismayed and yells: "How did we get into this fix? The old building is perfectly good, we really don't need air conditioning in Alaska; since we introduced those machines, things have gotten out of control!"

This is a typical story—one often told and perhaps even representing a situation often experienced. But just because such stories are told, and some people may interpret their experiences in this fashion, it doesn't follow that they have lost control to some autonomous technology that has taken over their company. What the tale allows us to see is that despite the fact that machines play a prominent role in the unfolding sequence of events, the major overlooked fact is that people often tend to forget the reasons for which they introduced a certain kind of tool or procedure. Instead of taking time to assess critically the impact of making further accommodations to the tools, possibly even concluding that it may be time to reexamine the whole situation, people often simply "go with the flow" and take what appears to be the course of least resistance. Still, from the fact that people sometimes tend to react to the circumstances of a situation in certain ways, perhaps accommodating a new procedure at first, rather than either replacing it either with another or eliminating it altogether, it does not follow that the procedure is autonomous.

A basic point we sometimes tend to forget is that *there is no getting rid of tools, written large*. Humanity making its way around in the world is humanity using tools of wide variety and complexity, e.g., hammers, automobiles, governments, electricity. The tools we invent to help us survive and go beyond are essential—perhaps even to the concept of humanity. It isn't as if we can remove tools altogether and continue without them. When we introduce an implement or a complex system, it is to help us achieve a goal. If we find that the device produces results or side effects in conflict with other goals and/or values, we may replace it or modify it. Whichever we choose, devices, tools, and systems remain with us; they are part of how we go about making our way in the world. What McDermott overlooked (when he spoke of how technologies become so ingrained in our procedures that in accommodating the requirements of the technology we lose our independence of action) was that it is the *perception, or lack of it, that people have of the usefulness of a new product that determines the extent to which they are willing to make concessions in its direction*. They may also lose sight of the goal that first guided their actions and, therefore, may react blindly to the circumstance with which they are now faced. But that is not to say that the product has "taken over." For nothing *in principle* rules our later modifications and, if necessary, replacements. What is required is that the individuals involved keep their objectives in mind and be strong enough to act in their own best interests.

Section 3. Common Sense

Phrased as I have put it, technology conceived of as humanity at work represents the results of the systematic application of common sense; common sense is how people first gain experience and then knowledge by acting on that experience. Nor should this result come as a surprise. Since, if we acknowledge that the concept of a tool lies at the commonsense heart of technology, and if we accept the rather obvious point that not all tools are physical tools, i.e., that there are conceptual tools, social tools, economic tools, etc., then it is not difficult to agree that knowledge is a tool, and if knowledge is constantly being updated, the tool is constantly being honed. In other words, if science produces knowledge, then the knowledge science produces is constantly being upgraded and changed by virtue of the impact of various other tools on the efforts of science to discover more and more about the world. Or to put it differently, quite aside from the resolution of the question of the independence or interdependence of technology and science, if science produces knowledge, and if that knowledge is sometimes used to develop tools that are used in the world, then what those tools produce should generate a form of knowledge that ought to have a bearing in turn on the original knowledge that produced the tools. In addition, it follows that what we do and how we do it is also constantly changing in the face of these developments, and that is as it should be. The bottom line is that, on this account, once a relation between a science and some tool or procedure is established, neither can lay further claims to autonomy—the interdependence is an essential aspect of the process of science itself.[2] But this point of view cannot be established only by *a priori* argument. We need to look at what actually goes on; and I have selected an historical case study to illustrate my points. This is not to say that the analysis of one historical example will settle the issue, but it should help clarify some matters.

Indeed, the case I want to look at, Galileo and the telescope, ought to help exhibit just the issues relevant to sorting out some of the confusions surrounding the interrelations between the development of science and the use of tools and systems of tools. Furthermore, there is a punch line. The general thesis, as already expressed, is that science and technology—where "technology" should now be read as tools, techniques, and systems of tools and techniques—where they interact at all, are mutually nurturing. There is also a caveat, to wit, in point of fact some technologies are science-independent, e.g., the roads of Rome. This is not to say that they are autonomous, since those technologies were responses to needs and goals also; just not the needs and goals of some scientific theory. And some science generates no technology, e.g., Aristotelian biology. The punch line is this: once that is said, something of a paradox emerges. For the history of science is the history of failed theories. But the failure of theory most often does not force a discarding of whatever technology that theory generated or was involved with, nor does the failure of the theory force the abandoning of the technology if a technology was responsible for that theory. To oversimplify: sciences come and go,

2. See Dewey 1929, for the development of a similar argument.

but their technologies remain. But oversimplification is what got us into trouble at the start, so a more accurate claim would be: scientific theories come and go, but some technologies with which they are in one way or another associated remain. It is also the case that some technologies associated with specific scientific descriptions disappear when they are replaced or superseded by new techniques.

But there is one sense in which the transient character of scientific theories becomes somewhat problematic. That is, if, as I put it earlier, technology is an integral part of science and partially responsible for changing the science, then the failure of the particular theories could be construed as a failure of the technology involved as well. This may in fact be true. But we should also emphasize our model, MT, in which technology is seen as a process of policy formation, implement-system implementation, assessment, and updating, which process functions at a variety of levels and with varying degrees of significance for technologies further up and down the line; e.g., the initial failure of the Hubble to produce clear pictures of the heavens did not spell disaster for the entire project. Goal-achieving activities are nested within one another and, as we shall see, as a matter of historical and physical accident the nesting will have different degrees of importance depending on the case. Thus placing the blame for a failed scientific theory on its associated technology once again oversimplifies the situation.

Section 4. Galileo and the Telescope

To illustrate some of the notions introduced here, let us turn to an examination of the development of the telescope by Galileo and its effect on some of the theoretical problems he faced in his efforts to show that Copernicus's theory was worthy of serious scientific consideration. As we shall see, the story is not a simple one, and the issue takes on an increasing degree of complexity as the tale proceeds.

To begin with, we need to be perfectly clear that Galileo did not begin his work on the telescope in order to prove anything about Copernicus. The full story of how Galileo came to construct his first telescope is clearly and succinctly put forth by Drake in his *Galileo at Work*. There, quoting from a number of Galileo's letters and published works, Drake makes it clear that Galileo was first drawn to the idea of constructing a telescope out of financial need. To summarize the account: in July 1609 Galileo was in poor health and, as always, if not nearly broke at least bothered by his lack of money. Having heard of the telescope, Galileo claims to have thought out the principles on which it worked by himself, "my basis being the theory of refraction" (as quoted in Drake 1978, 139). Drake acknowledges that there was no theory of refraction at the time, but excuses Galileo's claim on the grounds that this was not the first time that Galileo arrived at a correct result by reasoning from false premises. (Historians of the logic of discovery, take note.) Once having reconstructed the telescope, Galileo writes: "Now having known how useful this would be for maritime as well as land affairs, and seeing it desired by the Venetian government, I resolved on the 25th of this month [August] to appear in the College and make a free gift of it to his Lordship" (as quoted in Drake 1978, 141; Galileo's letter to his brother-in-law Benedetto Landucci). The result of this gift was the offer of a lifetime appointment with a nice salary increase from 520 to 1000 florins per year. What was unclear at the time, and later became the source of major annoyance on Galileo's part, was that along with the stipend came the provision that there was also to be no further increase for life. So he reinitiated his efforts, eventually successful, to return to Florence.

Now there are some problems here that need not delay us, but they ought to be mentioned in passing. How Galileo managed to reconstruct the telescope from just having heard reports of its existence in Holland remains something of a mystery. Galileo provides us with his own account of the reasoning he followed; but, as Drake notes, his description has been ridiculed by historians because, despite the fact that the telescope he constructed worked, he did not quite think it through correctly. Nevertheless, Drake's observation, that "the historical question of discovery (or in this case, rediscovery) relates to results, not to rigorous logic," seems to the point (Drake 1978, 140). Despite the fact that a telescope using two convex lenses can be made to exceed the power of one using a convex and a concave lens, the truth of the matter is that Galileo's telescope worked. On the other hand, this point about faulty reasoning leading to good results seems to tie into the paradoxical way in which technologies (thought of as artifacts of varying degrees of complexity and abstractness) emerge and remain with us. But more of this later.

We can now turn to the question of the impact of the telescope on Galileo's work. As he reports it, Galileo first turned his original eight-power telescope toward the moon in the presence of Cosimo, the Grand Duke of Florence. He and Cosimo apparently discussed the mountainous nature of the moon, and shortly after his return to Padua in late 1609, Galileo built a twenty-power telescope, apparently to confirm his original observations of the moon. He did so and then wrote to the Grand Duke's secretary to announce his results. So far then, Galileo has constructed the telescope for profit and is continuing to use it to advance his own position by courting Cosimo.

Galileo, never retiring about his work, continued to use the telescope and to make his new discoveries known through letters to close friends. Consequently, he also began to attract attention. But others such as Clavius now also had access to telescopes. That meant Galileo had to put his results before the public in order to establish his priority of discovery. Therefore, in March 1610 Galileo published *The Starry Messenger*, reporting his lunar observations as well as accounts of the Medicean stars and the hitherto unobserved density of the heavens. At this point controversy enters the picture. These reports of Galileo essentially challenge one of the fundamental assumptions of the Aristotelian theory of the nature of the heavenly sphere: its perfection and immutability. While the rotation of the Medicean stars around Jupiter can be shown to be compatible with both the Copernican and the Tychonian mathematical astronomies, it conflicts with the philosophical and metaphysical view that demands that the planets be carried about a stationary earth embedded in crystalline spheres. And to be clear about the way the battle lines were drawn, remember that Galileo's major opposition came primarily from the philosophers, not from the proto-scientists and other astronomers of his time.

The consequences of Galileo's telescopic observations were more far-reaching than even Copernicus's mathematical model. For the problems Copernicus set were problems in astronomical physics and, as such, had to do with meeting the observational restraints represented by detailed records of celestial activity. Galileo's results, however, and his further arguments concerning the lack of an absolute break between terrestrial and celestial phenomena, maintaining as they did the similarities between the moon and the earth, etc., forced the philosophers to the wall. It was the philosophers' theories that were being challenged when the immutability of the heavens was confronted with the Medicean stars, the phases of Venus, sunspots, and new comets. One might conclude, then, that this represented something akin to a radical Kuhnian paradigm switch.

Much has been written about the extent to which Kuhn's paradigm shifts and their purported likeness to Gestalt switches actually commit someone who experiences one to seeing a new and completely different world. But to see mountains on the moon in a universe in which celestial bodies are supposed to be perfectly smooth comes pretty close to making sense of what this extreme interpretation of Kuhn might mean. Prior to the introduction of the telescope, observations of the heavens, aside from providing inspiration for poets and lovers, were limited to supporting efforts to plot the movements of the planets against the rotation of the heavenly sphere. Furthermore, metaphysical considerations derived from Aristotle interfered with the conceptual possibility of learning much more, given the absence of alternatives. The one universally accepted tool that was employed in astronomical calculation was geometry, and its use was not predicated on any claims of realism for the mathematical models that were developed, another point derived from Aristotelian methodology. The acceptable problem for mathematical astronomy was to plot the relative positions of various celestial phenomena, not to try to explain them. Nor were astronomers expected to astound the world with new revelations about the population of the heavens, since that was assumed to be fixed and perfect. So whatever else astronomers were to do, it was not to discover new facts; there were not supposed to be any.

But the telescope revealed new facts. And for Galileo this meant that some way had to be found to accommodate them. Furthermore, to make the new telescopic findings acceptable, Galileo had to do more than merely let people look and see for themselves. The strategy he adopted was to link the telescopic data to something already secure in the minds of the community: geometry. This, however, was not as simple as it sounds. He had to build a case for extending geometry as a tool for physics, thereby releasing it from the restrictions under which it labored when used only as a modeling device for descriptive astronomy. In other words, Galileo had to advance the case of Archimedean mechanics. To this end he was forced to do two different things: (1) emphasize rigor in proof—extolling the virtues of geometry and decrying the lack of demonstrations by his opposition; and (2) de-emphasize the appeal to causes in providing explanations of physical phenomena (since abandoning the Aristotelian universe entailed abandoning the metaphysics of causes and teleology—without which the physics was empty).

Section 5. Geometry as a Technology

This is not the place to detail the actual way in which Galileo employed geometry to radicalize the notions of proof, explanation, and evidence.[3] Suffice it to say that he did and that it met with mixed success. The general maneuver was to begin by considering a problem of terrestrial physics, proceed to "draw a little picture," analyze the picture using the principles of Euclidean geometry, and (1) interpret the geometric proof in terrestrial terms, just as a logical positivist would interpret an axiomatic system *via* a "neutral" observation language, and then (2) extend the terrestrial interpretation to celestial phenomena. This is how he proceeded with his account of mountains on the moon, namely by establishing an analogy with terrestrial mountains. This process took place in stages. He first subjected the terrestrial phenomena to geometric analysis and then he extended that analysis to the features of the moon. Not all of Galileo's efforts at explanation using this method succeeded, e.g., his account of the tides. Nevertheless, the central role of geometry cannot be denied.

While Galileo used geometry for most of his career, it was not until he was forced to support publicly his more novel observations and hypotheses

3. I have worked on the topic; see Pitt 1978, 1982, 1986, 1991; as have McMullin 1968; Shea 1972; Wallace 1992; among others.

that we find in his writings the beginnings of what was eventually to become a very sophisticated methodological process. This procedure is most clearly evident in his last two works, the *Dialogue on the Two Chief World Systems* and his *Discourses on Two New Sciences*. But in the end the *geometric method* as employed by Galileo, or to put it more specifically, Galilean science, dies with Galileo. No one significant carried on his research program using his methods. Whatever impetus he gives to mathematics in science, his mathematics, geometry, very quickly gives way to Newton's calculus and the mathematics of the modern era.

Galileo's use of geometry was as much the employing of a technology conceived of as a tool/technique as was his use of the telescope. Furthermore, it represents the first major step toward the mathematization of what today we would call science. This much is commonplace. The challenging part comes in two sections. (1) The telescope was a new technology, whose introduction for primarily nonscientific reasons, i.e., money, was in fact science-independent, i.e., its invention by the Dutch was theory-independent. (The inventor, Hans Lipperhey, was a lens grinder; the invention was apparently the result of simply fooling around with a couple of lenses, the basic properties of which were known through Lipperhey's daily experience.) In many ways, the use of this new technology by Galileo can be held responsible for the extension of the *geometric method* as a radical method of supporting knowledge claims. (2) Geometry was also theory-independent. But, unlike the telescope, geometry was a very old technology. It was called upon to rescue, as it were, the new technology. It was a very different kind of technology from the telescope, being a method for providing justifications, i.e., proofs, of abstract conclusions regarding spatial relations, not a physical thing. Furthermore, despite the fact that this old technology was required to establish the viability of the new, the old was soon to become obsolete with respect to the justificatory role it was to play in science. That it was to be replaced also had nothing to do with any significant relation between the telescope and the development of the theory Newton outlined in his *Principia*. In other words, the telescope itself had little direct bearing on the development of the calculus, and yet it was the calculus that superseded geometry (but did not completely eliminate it) as the mathematical basis for scientific proof.

Section 6. Technology and the Dynamics of Change: Autonomy Socialized

If we try to sort it all out, the results are uncomfortable for standard views of technology and the growth of knowledge. The two technologies remain, the two sciences have been replaced. Furthermore, in one of the truly nice bits of irony that history reveals, one of the superceded technologies, geometry, after being replaced by a different kind of mathematical system for justificatory functions, experiences a resurrection in the nineteenth century and ends up playing a crucial role (but not a justificatory role) in the development of yet another physics, having been modified and expanded in the process.

Where is the autonomy here? Both Galileo's physics and the telescope, while capable of being viewed as independent products of one man's creative energy, can also be seen performing an intricate *pas de deux* of motivation and justification when the process of inquiry is examined. It is getting difficult to determine which view ought to take priority. A resolution of the problem might be found if we stop looking at the history and examine the concept of "autonomy" itself.

If we define "autonomous" as "free from influence in both its development and its use," then technology cannot be autonomous since it is inherently something used to accomplish specific goals. But what happens if we try to define "technology" so as to allow technology to have an impact on us as well as on our environment? Are we then committed to the view that, given a technology in use, there emerges from its use a self-propagating process outside the control of humankind? If (1) technology is a product, and (2) we do not add some additional properties to technology beyond its being a thing we manipulate, then (3) there is no reason why we should even begin to think of technology as not within our control.

In other words, we can talk of Galileo being forced to employ geometry and to develop novel methods of justification in order to defend his telescopic discoveries, for what sense does "forced" carry here? The telescope did not with logical necessity precipitate him headlong into battle. Much of what Galileo did to defend his claims and insure his priority of discovery was the product of his flamboyant personality. This was a man who loved fights and being in the public eye. How these features of Galileo's personality can be factored into the tool so as to make it appear that the tool itself is responsible for the action of the man is beyond serious consideration. Given the tool, we can plot its history. What that history amounts to is how it is used. How it is used is a complicated process, for it can entail more than intentional application of a device. "Use" may also mean "rely on," and it may be the case that what we rely on we take for granted, never giving thought to the cost. But this does not thereby entail that, in the absence of human deliberation, the tool by default acquires intentionality and, along with it, control of human affairs.

An alternative would be to endorse the idea that both the telescope and geometry used Galileo. This suggests a science fiction scenario in which as soon as any technology is used by a person, it "takes over" that individual. In the case of populations adopting constitutions that establish governments, all freedom of human action is lost since the government "takes over." Surely this amounts to a *reductio*. For the tool used to adopt government is reason. Is reason, too, going to be something sufficiently alien that we should fear it? The image really does become Mephistophelian enough that we ought to worry about the extent to which we have lost touch with reality.

Further, the existence of a technology does not entail that it will be used. We all know people who refuse to use computers today, not because they cannot, but simply because they feel more comfortable with the old technology of pen and paper. Surely we do not want to say that these individuals are controlled by pencil and paper. The decision to employ a certain means to an end requires thought, information, a determination of the nature and desirability of the end, assessment of the long- and short-term costs and benefits, as well as constant updating of the database. What if, in his declining years, our pen-and-pencil advocate changes his mind and opts for the computer, having decided that time is running out and he has too many things to finish by hand? Do we really want to say that the machine won out over man? Surely not; the man initiated the process that led to the machine, so why not include him in that process?

We are at the point where, in closing, we might ask: Why are we so quick to point to the machines and wag our finger? Well, the long and the short of it is that those who fear reified technology really fear men. It is not the machine that is frightening, but what some men will do with the machine; or, given the machine, what we fail to do by way of assessment and planning. It may be only a slogan, but there is a ring of truth to: "Guns don't kill, people do." There is no problem about the autonomy of technology. Pogo was right: "We have met the enemy and he is us" (Quoted in Kelly 1985, 114). The tools by themselves do nothing. That, I propose, is the only significant sense of autonomy you can find for technology.

CHAPTER SEVEN

Technology, Democracy, and Change

Among the many alleged "threats" technology is said to pose is a threat to democracy. The worry here is that technology, autonomous or unfettered or indiscriminately employed, will destroy our democratic institutions and the highly valued way of life we associate with them. There are many ways it is alleged this might occur, from the outright result of war brought about by a technologically trigger-happy military to the slow erosion of important democratic principles by the insidious invasion of specific devices. Typical examples include the problems caused by American television newscasters announcing East Coast results in a presidential election before the West Coast polls have closed, and the increasing ability of medical techniques to prolong life. With respect to premature announcements by television newscasters, the impact on the electoral process can be significant. We are not yet in a position to assess fully the social and psychological consequences not just of prolonging life, but of the impact of an increasingly large older population on the social fabric. One can, however, get some sense of the issue by reflecting on the potential the older population represents as a voting bloc and how this can impact on social relations and political agendas. Nevertheless, despite such examples, the general claim that technology presents a threat to democracy seems unwarranted. Instead, the situation is reversed; it is the continued development of new and improved means for making our way around in the world that are actually in danger from the misuse of the framework democracy provides.

Section 1. Innovation and Control

Those who see technology as a danger to democracy tend to base their perception on a number of different claims, but they essentially all boil down to one basic point: the size and pervasiveness of our technological infrastructure and the complexity of newly developing technologies are such that it is no longer viable for individuals rationally to control events in their society, not to mention the further development of technology. The more this complex structure develops, so the story goes, the more we are disenfranchised (see the discussion of Langdon Winner in Chapter 5). The evidence for this view is seen in an increased reliance on "experts" to inform us of the advantages and disadvantages of pursuing this or that development, and in the growth of a class of professional managers, "technocrats." This argument, however, is confused. Not only is there evidence that we *can* control the spread of large-scale technologies—witness the recent decision to dismantle the Long Island nuclear reactor—but the assumption does not support the conclusion. If we assume for the moment that we are in fact relying more and more on experts and technocrats, from this it does not follow that democracy is in any way threatened. There *are* threats to certain entrenched *value systems* that emerge with any form of social change. But from the fact that innovation may end up encouraging social change that either manifests or contributes to value change, it doesn't follow that democracy, as a basis of government, changes. We will examine this point in more detail later. But first, some preliminaries.

Some aspects of democracy do change with innovation, that is true. Consider, for example, the introduction of mechanical voting machines to register and tally votes automatically. That we no longer write out the names of the candidates we are voting for, or check a box next to their names on a preprinted ballot (a technological innovation?!), but rather rely on a mechanical contrivance doesn't mean that we have shaken the foundations of our democracy, although when these machines were first proposed, such arguments may have been advanced. Clearly, simple innovations like voting machines are not threatening. Moreover, they are not the type of "technology" at issue. But the existence of such cases as the voting machine suggests a need for some means of distinguishing relevant from irrelevant forms of technology as they bear on the issue of constituting a threat to democracy. With the need for such a demarcation criterion in mind, I suggest the following: *The type of technology that can legitimately be said to constitute a threat to democracy is one that challenges values that are perceived by some significant portion of the population as essential to their conception of the good life*. For example, the introduction of centralized computerized government records was opposed because it was contended that the resulting system would constitute a threat to the privacy of individuals. Whether or not such a system would constitute such a threat is not the point. It is important to note, however, that if it were a threat, it would not be a threat to democracy *per se*; it might be a threat to a particular conception of the ideal social environment, but that remains to be seen.[1]

But even if we accept the idea that those technologies that pose the kind of threat that might be marked as a threat to democracy do so by threatening some entrenched set of values, it doesn't follow that if such a set of values is threatened, then so is democracy. Thus it is a long way from acknowledging that the value of personal privacy is at risk to the claim that democracy is at risk. It is not even clear that personal privacy *is* a value crucial to democracy. If it is, it is a convenient one carefully circumscribed, since we do not accord our public officials or state prisoners such a right. It seems to be the sort of right one trots out when one feels personally threatened. But a personal threat is not a threat to democracy as such.

The view that technology threatens democracy is difficult to understand. This can be attributed to two different sorts of reasons. To begin with, there are a number of difficulties in the core of the concept of democracy itself. But, problems or not, it is hard to see how a political system can be threatened by anything other than its own incoherence. On the other hand, what is clear is that there are many respects in which both new scientific ideas and inventions and innovations are *perceived* as threatening, and that *under the veil of that perception, the techniques of democracy are employed against both science and technology*. Now this is something that *is* dangerous. For science is essentially a knowledge-generating process and, as such, it is used as a tool in our efforts—call that technology—to improve the human condition. If either (or both) our efforts to improve our lot or the development of improved tools we need to further that end are disrupted, then our future ability to improve humanity's lot is endangered. That being the case, I suggest that it is not democracy that is being threatened, but rather the development and use of new technology and, by extension, humanity. In effect, under the banner of preserving democracy, we can perversely undermine it. Not that this is done with the intent of undermining democracy. Rather, the motivation for attacks on new scientific initiatives and technological innovations seems to derive, and properly so, from the perception that such developments threaten a privileged set of values (meaning by that the values of whoever is objecting). As such then, the situation

1. Some technologies are also said to constitute threats to human existence, e.g., nuclear energy plants and the Strategic Defense Initiative (SDI). The extent to which they do is arguable and indeed has been argued at length. These are not the kinds of threats with which we are concerned. But, once again, as examples, nuclear energy and SDI represent different categories. Where nuclear energy is concerned, constant monitoring of the development of the technology and the results of its implementation can result in changes that might correct problems as they develop. With SDI, on the other hand, it appears to be an all-or-nothing situation: either it works or it doesn't, and if it doesn't, it's going to be too late to do anything about it.

we are examining is really best described as representing a *conflict of values*. On the one hand, we have what I will call the values of technology, meaning the values of humanity at work, and, on the other, there are the values of particular sociopolitical groups bound together by attachment to a particular lifestyle. The difference between these two groups may seem slim, but it is significant. The values of humanity at work cannot be spelled out *a priori*. They emerge and change as people wrestle with the problem of what they see at that time as improving the human condition. The values of the second group are set in stone. They are taken as given and they are not subject to change. Change is what threatens these values.

To refer to the values of technology is not to make an ontological claim to the effect that there exists something called "technology" that by itself has a set of values. Rather, in making our way around in the world we are constantly faced with decisions. All such decisions involve choices, and all choice proceeds against a background of justificatory values. We have already noted that not all values are moral values. There are, for example, aesthetic values, religious values, and cognitive values. The values of technology, where technology is conceived as primarily an epistemic activity, are fundamentally cognitive or epistemic values. That is, *the values of technology are those values that govern the process of acquiring, testing, and using knowledge* for the purpose of changing the natural and social environment for the better. Such values include the notion of the truth of a statement, the reliability of data, the importance of testing, and the recognition that all knowledge claims are subject to revision in the light of new information.

Furthermore, these values are evolving over time, as is the very nature of human knowledge. In the light of the *Challenger* disaster, we have witnessed the value of giving reports of potential danger higher priority. And we are beginning to watch the emergence of the importance of ethical considerations in the formulation of the professional identity of engineers. This is sure to produce situations in which there is yet another conflict between ethical and epistemic values. Given the changing character of our knowledge as well as the changing structures within which we conduct our work due to the influence of new tools and structures, it becomes very difficult to isolate technology as such. To make matters more difficult, when discussing these issues, we tend to use general terms whose meanings evaporate when we look at specific cases. To put it as bluntly as possible, having already argued that there is no such thing as technology *simpliciter*, nor is there such a thing as *science*, I now propose that there is no such thing as *democracy*. That being the case, it becomes increasingly unproductive to try to capture the flavor of disputes about the potential dangers of technology in general terms. But when we look at specific cases, we see that what is initially characterized as a threat can only be seen as such if we understand how a particular innovation challenges a specific value. Thus, for the most part, general talk of technology and democracy illuminates very little.

Nevertheless, some general talk about technology is necessary by way of laying out some grounds for coming to grips with some of the relationships among science, technology, and democracy. To do this we need only reflect on the discussion so far. In the first sections of this chapter we have essentially been talking about technology as the *complex of activities* that combine as the activities of humanity, each impinging on another and rebounding to affect others. If we unpack that idea, we see that the exploration and manipulation of our world to achieve specific goals and the continuous updating of what we know based on that process are *epistemically prior* to whatever theoretical conceptualization we have developed as part of the process. That is, *technology is epistemically prior to science because when technology is conceived of as humanity at work, action precedes theorizing*. Theorizing grows out of action by way of reflecting on the success or failure of the action. As human beings, our major concern is to act, changing our environment to meet our goals and needs. We act first, learn from our actions, and then try again. This is the sense of the pragmatic ideal that knowledge is for the sake of action (Lewis 1946; CP: 5.384-387). It is not just that we need to know before we can act. *The need to know derives from the need to act better*. Knowledge, including science, is secondary to acting. In other words, knowledge, i.e., science, is at most another tool. But from the fact that technology precedes science in the order of being, this does not mean that technology is only one *thing*. Rather, it is the feature common to all human behavior. In seeking to obtain our goals, we develop and then change a variety of structures, organizations, and devices to assist us. That is the nature of the beast: to grow in the context of learning to survive.

If there is no one thing that can be identified as technology, beyond the sum of human activity, what about "democracy"? Surely no one would attempt to argue that there is some one political ideal that incorporates democracy. At best, when we talk about democracy we mean something like the idea that in whatever political unit you are referring to—state, university department, or family—"the people" have access by means of voting to the decision-making process whose end products affect their well-being. No particular arrangement seems superior to any other. The many forms democracy has assumed do not suggest that one instantiation is necessarily better than another. It is not clear that the New England town meeting is better than the representative federalism employed at the national level in the United States. Nor is it clear that majority rule is the key to democracy. What is essential is that each citizen, member, appropriately designated participant, etc., be allowed to express his or her opinion by voting, and that the final decision be a function of that vote in some manner or other. Now, it is not at all clear how technology, characterized either as the sum of human activity or, more conventionally, as some structure of tools in the context of an enterprise designed to exploit some aspect of our society or environment, can threaten democracy. On the surface, it appears that the only threat to democracy is some action that deprives the individual of his or her vote.

It might be objected, however, that things are not that black-and-white. Surely, it will be argued, it is not just a case of whether or not you have a vote. The real issue is whether or not you can cast an *informed* vote. The threat to democracy posed by the increasing size, complexity, and expense of large-scale technological undertakings is that no ordinary citizen can understand the scope and consequences of the enterprise, hence having a vote means nothing since you can't understand what you are voting for.

One response to this worry is to introduce the expert who is supposed to know enough to cut through the morass of technical issues and give us the bottom line. The ordinary citizen need not know everything. What the citizen must be able to do is to evaluate critically the arguments placed in front of him by those who are experts in the areas under discussion. But, it is countered, this is exactly the problem; the citizen must rely on the expert and hence can be manipulated. But this is no more a problem than the problem we face with representative democracy. We elect representatives to govern us, and they may not always vote as we would, or they may be corrupt. The fact that each citizen does not vote on all issues pertaining to the national interest does not mean we have abandoned democracy, nor does it mean we are being manipulated in some intrinsically evil fashion. Furthermore, it is not clear that we really want to vote on each issue. The irony of the situation here is that innovation has provided us with the means to achieve something close to a New England town meeting on a national scale through the use of computers, television, and telephones, but it is unlikely that we will take up the option. One conclusion to be drawn from these observations is that if having the vote is to be considered a crucial part of democracy, then what it means to have a vote needs to be spelled out in some detail. I suspect that when all is said and done, the value of the vote is an *in principle* concept.

Reliance on experts does not clearly represent a threat for two reasons. The interesting thing about experts is that they will ultimately be proven right or wrong, and when that happens, the results will be fed back into our evaluative system, upgrading our database, giving us new and better informed grounds for making the next decision.

The second reason why relying on experts is not a threat is that for some situations there are no experts, despite proclamations to the contrary. There are a number of situations in which the appropriate knowledge is simply not available. Once we realize that, the illusion of experts controlling our decision-making should be weakened. One striking example that comes to mind here is the use of experts in the form of psychologists and psychiatrists to determine the sanity or insanity of a defendant in a legal situation. When two sides to a conflict are able to muster equal numbers of experts to support opposing views, we should conclude that this is one of those situations in which the appropriate knowledge is not available. The problem then becomes how to determine what the appropriate knowledge might be. Under those circumstances, it would also seem rational to suspend decision-making until the issue of relevant knowledge was resolved. Here the important thing is that we do not need to be experts to realize that the "experts" don't know what they are talking about. If there is a danger here, it lies in finding ourselves in situations in which either the appropriate knowledge is in principle unobtainable, or its acquisition requires a timescale that renders the immediate problem insoluble. If the appropriate knowledge is in principle unobtainable, then we have to rethink the entire enterprise. If the timescale is too long, we should probably attempt to break the problem down into more refined components. Whichever the situation, it should also be remembered that there is an important sense in which all of our knowledge is inadequate to the job of making decisions about which we can be totally confident. If we accept the view that knowledge is always undergoing revision, then the justification for all of our past decisions is continually being undermined. In short, while we are more and less sure of many things we claim to know, we can be certain of nothing. This means that all decisions in principle are made without the appropriate knowledge. It is a question of degrees of confidence, which is not an indicator of the accuracy of data.

Assume for the moment that the argument so far is coherent. A problem remains. It might be argued that the real issue is not what we have been attending to up to this point. The real issue is that, because of the complexity of contemporary technological innovations and their long-term consequences, once we get started we can't *afford* to abandon a course we have taken on the basis of expert advice when that advice turns out to be incorrect. Given the costs, we end up trying to correct our mistakes in a piecemeal fashion, spiraling deeper and deeper into the clutches of a nightmare situation from which we cannot extricate ourselves. Having committed to one of these big projects, the chances of backing out once we discover our mistake is zero. The momentum of technology, so it is argued, lumbers on over us despite individual protests. Once we are caught up in the momentum of technology, the individual vote means nothing. *That* is the threat to democracy. But that sounds like claims we have already examined in the context of looking at the notion that technology is autonomous.

In response, therefore, we need only observe that the unwillingness to cut losses, when it is discovered that a given action has produced bad results, is a failure of people, not a property of technology writ large. There is no momentum to technology. At each juncture people must make decisions to continue. Yes, it may be expensive to say "no more." But it would also be irrational to continue once we have discovered our mistake. The decision to dismantle the Long Island nuclear plant was brave and expensive. But it is just the sort of decision that exemplifies our ability to control our own destiny. Likewise, the decision to abandon the superconducting supercollider project. To blame "technology" for irrational or cowardly behavior on the part of those given responsibility for leadership is silly. Technology, viewed conventionally, isn't the culprit; we are. That being the case, we can identify one threat to democracy: an irrational (in the CPR sense) people.

Earlier we characterized the problem of the alleged threat by technology to democracy as a conflict of values, the values of technology versus those of some sociopolitical group operating in a democratic framework. To make sense of this claim, we embarked on a preliminary analysis of what we mean by "technology" and "democracy," with an eye to isolating the values in question. It is important to realize at this point that democracy as such is close to being value-neutral. That is, to endorse democracy is not necessarily to endorse the view that individuals have a right to privacy, or any other right for that matter. Consider even the right to vote. Not all individuals are accorded that right, although in a society that proclaims itself a democracy, we often witness the struggle for that right on the part of the disenfranchised, as in South Africa. In the United States there is an age limit for voting, and federal felons are deprived of the right to vote. These aspects of our governance system do not represent any intrinsic values of democracy *per se*. What they do represent is the result of a particular set of culturally bound experiences. For example, in defense of the voting age, it is argued that experience has shown us that one must acquire a certain level of maturity, experience, etc., before one can be expected to vote rationally. Of course we have no evidence that "adults" do any better at this than juveniles, but that is the argument.

Are there any values that democracy represents as such? It is not clear that there are, but it is probably safe to say that if there is one, it is that the people, however that is construed, ought to hold ultimate authority, however that is construed. The forms this value can take seem to be infinite. To begin with, it is not clear who "the people" are. The people can be the full population of a determined order, or some segment. How this is resolved seems as much a matter of force or circumstance as anything else. The sense in which the people hold authority is equally vague. But in the long run, it must mean something like the idea that the exercise of power without the legitimation of the people being governed is the rejection of the value of democracy. And, it should be noted, there are good reasons for rejecting democracy in some cases. It is not self-evident that democracy is necessarily the best form of social arrangement in all circumstances. That is, it is not at all clear that every social arrangement is best governed by democratic means; take a corporation, for example; or consider a university or the army. It may even be the case that democracy may not be the best form of government to develop and control technologies.

So far, then, we have characterized the values of technology as epistemic in nature, and we have discovered that if there are values implicit in the concept of democracy, there is at most one political value the adoption of which entails endorsing the legitimation function of the people being governed. This now puts us in a position to examine the earlier claims to the effect that we can't make sense of the idea that technology is a threat to democracy, and that the real situation is one of value conflict between the values of technology and the values of some group functioning within a democratic framework.

To put the issue in its most basic terms, can the search for the best basis for future action, i.e., technology conceived of as an epistemic activity, constitute a threat to the legitimation of a government by the governed? Yes, there is a possible positive answer to that question, but only one, and upon examination we will see that it ultimately makes no sense. Technology can threaten democracy if, in the ongoing search for knowledge based on experience, it is discovered that democracy is not the best way to acquire the most effective basis for achieving our goals. We would have arrived at that decision only after having acquired sufficient evidence to that effect. That being the case, it would be irrational to continue to endorse democracy. But for it to be a rational decision to reject democracy, it would have to be based on the knowledge that *all* forms of democracy have failed. It would not be enough to show that in this or that form of democracy, the goals of the society cannot be met. Furthermore, it is not clear that just showing that the goals of a society cannot be met is sufficient to reject democracy. For it might be the case (and probably is) that the goals of the society, to the extent we can talk about them, are incoherent, in which case it would not be democracy's fault they cannot be met. Finally, one of the things we have learned over time is that what we now know is bound to change and as what we know changes, our goals change. It would therefore follow that it would be unreasonable to reject democracy on the grounds that, given what we know *now*, our current and possibly incoherent goals can't be met.[2]

---

2. Just as the conclusion that the current arrangements of democracy fail cannot force us to give up on democracy, such a conclusion cannot force us completely to reject any form of technology or the technological imperative, i.e., the need to develop new technologies. For example, take the case of nuclear energy. What we have learned from our experience is that, given the current state of affairs,

In 1976 the City Council of Cambridge, Massachusetts, following some rather noisy public hearings, voted a moratorium on the construction of a Harvard University laboratory to be used for recombinant DNA experiments. This decision was the result of public outcry, contributed to in part by a general state of disarray within the scientific community. The techniques of manipulating DNA were new; not much was known about their ramifications, and there were other legitimate factors to consider. There was the questionable state of the building in which the P3 lab was to be constructed.[3] There are good grounds for thinking it was not safe. The situation was complicated, but it is clearly an example of a small social group affecting scientific research. The question here is not whether the protesters were justified in so doing or not. That they, at least temporarily, accomplished their goal through democratic means is a clear example of the kind of threat the use of democratic institutions poses to science in the traditional sense and technology in my sense. The DNA case is a good one for our purposes because research into recombinant DNA clearly involves both the techniques of recombination and the scientific quest for knowledge, each reinforcing and restructuring the other. This represents a good example of what I have been referring to as technology as an epistemic enterprise developing knowledge in the context of manipulating the world. Thus, while the challenge here is nominally directed at scientific research, the epistemic process of technology is what is clearly being affected.

Among the many interesting features of this case are the following. Looked at in one way, the situation is more frightening than appears at first glance. This is not merely a case of a small community, fearful for its safety, taking steps to insure the general welfare. Instead, it is an example of a small group using local democratic institutions to thwart work that is essentially global in character as well as in its ramifications. Unlike the City of Cambridge, the scientific community is worldwide. This demonstration of local democratic power shows the way to further uses elsewhere and in other circumstances. So while democracy as such does not pose a threat to technology, specific uses of specific democratic institutions can threaten general scientific and technological developments. It should be emphasized, however, that there was good reason to be afraid of building such a facility at that time in that place.

---

our ability to manage nuclear energy seems inadequate. We should also note that while that seems to be true of America, it is not true of France. But as the greenhouse effect (if there is a greenhouse effect) worsens, and as our knowledge of materials and matter improves, we may find that returning to nuclear energy is the best way to go.

3. "P3" refers to the level of security of containment of a laboratory.

People do in fact become frightened when faced with an unknown with potentially far-reaching effects. Furthermore, this is a natural and understandable reaction. We should not forget that this is the age of Big Technology. Memories of the use of the atomic bomb in World War II are still fresh and they still produce a powerful impact. We are and ought to be afraid of what we have the power to do and the scale of the consequences.

However, we should also reflect on the fact that we also know more about how to recognize potential dangers and we know how to proceed accordingly. It would be a shame, from an epistemic point of view, if fear resulted in stasis. What we must learn how to do is (1) look for potential dangers associated with various new initiatives; (2) learn to suspect our methods of assessment and to subject them to rigorous evaluation and reevaluation; (3) be willing to slow down if our assessment procedures reveal a confused conclusion. But it doesn't follow that every objection raised has merit.

Consider a second example. Jeremy Rifkin has made a career out of using the power of the courts to interfere with scientific experimentation. He defends his actions on the grounds of the need for further knowledge of the consequences of such experimentation. In this case, it is not a community we are discussing, but one person who has managed to use one aspect of this democratic framework, i.e., the courts, to interfere with the development of knowledge. Rifkin has most recently suggested that we proceed with DNA experimentation using the Hippocratic Oath as our guide. "The rule of thumb ought to be the Hippocratic Oath: First, do no harm" (Quoted in McDonald, Galarza, Rose 1998, 93). This is a deceptively simple approach to these complicated issues. The do-no-harm injunction is an open-ended invitation to inaction since it is always possible to ask the question: are you sure? As we have seen, knowledge is not certain, hence if we don't know for sure what the consequences will be, it would follow that we shouldn't go ahead with this or that project.

In response, let me suggest that the status quo, as we have seen, is no defense against doing no harm. Let us be proactive and engaged. This means that, instead of appeals to the Hippocratic Oath, we should make our maxim for action: Do Good. To sit immobilized in the face of danger because one might do harm is to guarantee disaster. We must act or we will fail, no matter what. Action brings with it the risk of failure. But only then can we learn and improve on the possibility of success for the next time.

A third situation worth considering is the destructive use of public political power to undermine the value of the search for knowledge. Senator William Proxmire's Golden Fleece awards threaten with ridicule research and innovation and the power they can muster for good. Offered under the pretense of public guardianship, his technique of holding up esoteric and sometimes overtly laughable projects as examples of public waste have the unfortunate feature that he never gives the opposite side. That is, he doesn't ask for, nor does he provide, any account of the possible long-range benefits of such research. It doesn't matter that such a defense could be given even at some later time. The very public act of deliberately humiliating scientific research, the perpetrator wrapped in the toga of public defender, suggests a frame of mind that assumes that the well-being of a society owes nothing to the careful and wide-ranging search for knowledge. This is not to deny the value of healthy skepticism and the importance of public scrutiny in all aspects of the public domain. It is, rather, an example of the kind of threat that can be mounted against science and technology in a democratic framework. There are others as well. Consider the effect of massively funded government projects on what ideally should be an unhampered examination of the truth of developing theories about the universe. When we recognize that most American scientists conduct their research in universities that rely heavily on overhead from research funds, we can begin to see how the existence of a large public bankroll can corrupt investigators by influencing their choice of research topics.

In short, the attack on technological dimensions of social and epistemic activity through the democratic framework is varied and constant. That it has not been successful in completely undermining these activities can be attributable to two factors. There is, first, the enormous success of the results of research, innovation, and invention in providing us with the means continually to upgrade our quality of life. But should there be a massive DNA mistake, for example, I suspect that recombinant DNA research would be curtailed, and that would be accomplished using all the possible vehicles our democratic society allows. Look at what happened to the development of commercial nuclear energy following the Three Mile Island accident. It is probably only because of its military value that we are still funding any nuclear research.

The second reason the tool- and technique-based dimensions of society have not been undermined is that, just as there are those who attack "technology," there are those who support it using the same techniques as their opponents. And I suspect that many of the attacks are as much a result of objections to the misuse of local democratic frameworks to support technologies as they are an outgrowth of objections to the technologies themselves. After all, it was public money that was being spent on President Reagan's Strategic Defense Initiative. We should not be surprised when people object to spending that money because doing so cancels out their own political agendas. One can only sympathize with those who wish to use public funds to feed the poor but whose lobbying efforts to that end are overwhelmed by rich and powerful economically motivated interests connected to the defense establishment or the space program. And while, for example, humanists have a right to object to having their projects overlooked in favor of those projects backed by economically strong political lobbies, we must caution against identifying all technological issues as opposing humanistic issues. Nevertheless, technology does have its friends who abuse the democratic framework just as technology's opponents do.

In the cases we have been examining, the issue has been the abuse by individuals of a framework that is generally concerned with providing the means to improve the common good. There is a final abuse of the system that represents the greatest threat to both science and to the use and development of new techniques and tools. This threat comes in the form of a specific effort to terminate research in a particular area because it, or its consequences, are in conflict with the values of a small group within the society, values that are not shared by the larger society. As an example, consider the efforts of those opposed to *in vitro* fertilization. Or, to put the emphasis in a slightly different place, those who seek a constitutional amendment banning abortion are attempting to use democratic means to establish a value that has less-than-universal acceptance and is in conflict with another value that has been developing over the past three decades, the right of a woman to control her own body. Now the point of this example is to show that what is at issue here is not abortion, but the idea that democratic means can be used to thwart as well as enhance the efforts of individuals seeking to improve the quality of their lives. What people do is generally done for the sake of what they perceive to be an improvement in their own circumstances and sometimes the circumstances of others as well. That there should be disagreement over what constitutes an improvement is no surprise. That is what value conflict is all about. But when the challenge is not to the specific improvement, but to the very process of seeking improvements, then we have a problem. Right now, the same process that allows the possibility of a constitutional amendment banning abortion can also be used to stop specific forms of scientific and technological research. Likewise, the same methods can be used to define as science something that does not meet the standards of science, i.e., creationism. It is a worrisome situation and one for which there is no easy solution. At best one can hope for a reasoning society, one in which the long-term value of science and of our technologies can be appreciated even by those who recognize the threat new knowledge and new techniques present to entrenched values.

Section 2. The Race between Technology and Morality

One last issue needs to be addressed. The issue here is the often-heard lament to the effect that our technologies are developing faster than our moral abilities to handle them. What this seems to mean is that new technological developments raise new moral questions for which we are unprepared. For example, there are lots of moral questions associated with the techniques of recombinant DNA technology. The questions range from worrying about the appropriateness of actually modifying human DNA to produce individuals better fitted for certain tasks, to worries about the appropriateness of releasing artificially engineered microbes into the environment. And, no doubt about it, these are legitimate worries. What is not clear, however, is the legitimacy of one of the conclusions that have been drawn, namely, that we should stop developing these advanced techniques. In this case, the reason the conclusion seems unwarranted is that it is based on two assumptions. First, that we need time to develop our moral theories and conceptual equipment so as to be in a better position to decide what to do. Second, that our current moral theories are inadequate.

The key to dealing with this concern is to realize that ever since a human first used a tree limb to batter or kill another creature, our technologies have outpaced our moral theories. Further, it is not at all clear that moral theorists have been able to resolve even the simple problems, such as the morality of war. To demand that we delay technological developments until we have adequate moral theories to handle new developments seems not merely reactionary, but naive. No one will deny that some new technological developments raise moral issues. What is not clear is whether there are any moral theories that can be produced *a priori* that will be able to handle these problems. The relation between morality and our technologies ought not to be one of censorship in ignorance. Knowledge is one of the fundamental values of our society. How to use that knowledge has been and remains one of our great problems. What we have discovered is that the use of new knowledge is best decided by public debate and experience. That means that the moral assessment of new developments will also follow behind. To do otherwise is to assume that moral theories have a greater possibility for certainty than do scientific theories, something for which we have abundant evidence to the contrary.

Section 3. Conflicts of Values and Technological Change

In the last several chapters it has been argued that there are some things we ought not to say about technology because not only are they false in some instances, but that in all cases, by characterizing technological issues in these ways, we greatly reduce our ability to integrate these worries into our larger philosophical enterprise. Thus not only is it not the case that technology is autonomous or ideologically value-laden, but conceiving of technology as having these attributes puts us into the role of social critic without justification. Likewise, characterizing technology in ideological terms or urging restraint on technological development because of our inability to provide moral grounds for action places us outside the context in which reasoned argument about the merits of certain forms of development can take place. In short, if we are to engage our technologies, we must eliminate artificial conceptual barriers that may make us sound authoritative but that really restrict our field of action. But having eliminated these barriers does not automatically make it possible to deal reasonably with the kinds of value issues technologies raise. In particular, conflicts of values over new technologies lead us to questions about technological change.

There are two different aspects of technological change that seem to be appropriate for examination here. The first involves a recasting of the problem of technological change in the light of our definition of technology as humanity at work. A consequence of viewing technological change in the light of this approach is the opportunity to develop a justification for addressing some aspects of technological change as social issues, where no such justification has been available before. This, I will argue, is one of the appropriate places for the social critic to enter the discussion of the impact of technological developments on society.

The second aspect of technological change to be addressed here concerns the roles of the many technologies involved in scientific change. In this case, in keeping with our approach in earlier chapters, "Technological Change" is treated as a counterpart concept to "Scientific change." Current accounts of scientific change treat the issue as an exercise in the logical reconstruction of the history of ideas: How does one theory come to replace another? comes to mean nothing more than: What are the logical conditions under which it is rational to reject or accept a theory? The social, political, and economic contexts of science are ignored in these accounts, as are the contributions of developments in the support systems in which science is embedded, what I call *the technological infrastructure of science*. The consequence of considering these factors is that the so-called traditional treatment of the relationship between *scientific* and *technological* is reversed. Instead of seeing various technologies as spin-offs of scientific discoveries or as the handmaidens to scientific investigations, whatever that may mean, I propose to cast scientific changes as a result of the mutual interactions among scientists working with specific sets of ideas and the various technologies they use and rely on to make those ideas comprehensible and testable in a social context.

Section 4. Changing the Way Humanity Works

If we take our definition of "technology" as "humanity at work," then "technological change" comes out meaning something like "changing the way humanity works." It is in this context especially that the definition of technology as humanity at work clearly comes into its own. For when we speak of technological change in ordinary life, we do not mean just new gadgets. We talk about technological change in terms of the difference these new gadgets and ways of doing things make in our lives. More often than not, we change the way we work by introducing new ways of performing tasks. This can be accomplished sometimes simply by rearranging the way we do things. This is what "restructuring" was supposed to mean when it was introduced to the business world.[4] Sometimes we find new support systems to do the job previously done by other systems, as in the rise of alternative mail delivery systems to the official national postal service. And sometimes we introduce new gadgets, new things, to replace the old things that we used to use to do the job. A walk down the hall of an office building reveals in interesting ways this sort of change. We no longer hear the sounds we once heard in a business office: the clicking of typewriter keys, the sound of the bell indicating that you needed to return the carriage to its starting place, the ring of a telephone coming from metal clangors. Today, instead, there are the sounds of computer keyboards, printers, electronically generated telephone "rings," etc. And while the same tasks are performed, e.g., letters are written and reports compiled, the tools we use to do them have changed.

If this were all there was to technological change, I suspect there would be little to talk about. But clearly there is more going on here than meets the eye. We are not just replacing one tool with another in these situations. There are consequences associated with the changes. First, there are the economic effects for the manufacturers of the old tools. Typewriter manufacturers either go out of business or begin to develop and market computers. In a case where new gadgets replace old ones and, because of whatever factors are involved, the old company closes, there are social consequences as well: people lose their jobs, whole communities may be affected, there is the physical and psychological disruption associated with finding new employment, etc. And while these are clearly negatives, they are not enough to explain the sometimes fiercely negative reactions to the promise (threat?) of technological change.

---

4. Unfortunately, "restructuring" has come to mean "downsizing." It was not originally intended to lead business in that direction. In its initial versions, "restructuring" meant just that: rethinking the way things were done in an effort to be more efficient. That, however, requires work. Managers incapable of thinking through such a difficult process substituted the idea of more work for fewer people for the innovative techniques that restructuring calls for. At this time it seems that the problems of trying to get fewer people to do more work are coming to the fore, and some managers are, finally, rethinking the wisdom of earlier actions. When we look back on this episode in the history of business, the lesson may be that it takes a series of cycles for innovative concepts finally to come to function in the way they were originally intended.

The problems associated with changing the way we work come not from what we know will happen, but from what we don't know, i.e., *the unintended consequences*. The problem comes not from changing the way people work. It is both reasonable and desirable to improve our work situations—and the assumption, right or wrong, is always that the changes coming from using new tools or support systems are for the better. Rather, the primary problem comes from not knowing what kinds of *additional* consequences the introduction of new tools or the use of new support systems will bring. It appears to be the case that while even those opposed to technological changes may agree that a particular change will improve the immediate circumstances, they will, at the same time assume, rightly or wrongly, that all the unintended consequences will be negative. It is equally interesting that we seem better at predicting negative unintended consequences than at envisioning positive ones, or maybe it is just that the negative scenarios get (and make) better press. For example, at the time people were objecting to the prospect of centralizing and computerizing Social Security records on the grounds of privacy issues, no one was making much of a fuss about the World Wide Web and the boon it would be to research and to business communication.

Well, what are we to do about unintended consequences? Before dealing directly with technological change in the sense of changing how humanity works, we must consider at least one significant point. Unintended consequences are not only or always the result of technological change. Sometimes, there are unintended consequences that stem from continuing to do the same old thing the same old way. For example, continuing to dump animal waste into local waters can produce a toxic buildup in the water system, as it did recently in the Netherlands. Continuing unabatedly to produce children results in overpopulation, famine, misery, disease, poverty, and death. Continuing to expose one's body to the sun in the summer, e.g., tanning, ultimately produces weathered and leathery skin and increases the risk of skin cancer.

So it seems unreasonable to object to technological change solely on the grounds of not knowing all the potentially negative consequences that introducing new tools and ways of doing things will produce. We never know all the future consequences of any action. If knowing all the consequences were a necessary condition for acting, there would be no action of any sort. We couldn't even continue to do the things we used to do once we realized that continuing to do things the same old way can also have detrimental results where once the results were beneficial.

The justifiable basis for the resistance to technological change comes, rather, from not knowing what the effects of these changes will have on our *values*, on those features of our worldview that capture what we currently endorse as the most desired states of affairs. It isn't the direct effects themselves of technological change that we are suspicious of, but rather the effects of the effects of the introduction of innovations on those ideas, goals, and ways of living we claim to prefer. Further, I propose that we tend to resist technological changes for two reasons: (1) it is easier to envision threats to our preferred way of living than to consider how to improve it; and (2) despite the fact that we tend to assume that our way of living is the preferred way of living, i.e., that our values are the preferred, even "correct," values, since values are in principle unjustifiable by themselves, lacking a clear-cut and obvious justification for our own values, we realize the indefensible position for which we are posturing. The fact of the matter is that if we could justify our values conclusively, there would be no more than one religion, if that, and no quarrels over what is the right thing to do. And yes, it is true that just because we haven't found such a justification, there could be one around the corner—but I doubt it.

I have, in effect, been arguing that the perceived problems associated with technological change, where technological change means introducing new tools and new support systems, are at bottom problems of value change. And these are not the problems of merely having to figure out how to adjust to a new way of dealing with the world. The problems of this kind are exacerbated by a lack of knowledge of the consequences of introducing these changes for the way we live and the way we see the world. Often we are not confronted with a change that gives us a clear-cut choice between two sets of specifiable values; rather, in an attempt to project the effects of introducing new technologies, we construct situations in which our current values might be threatened by sets of unknown consequences. Further, we project the negative scenarios because it is easier than trying to imagine realistic positive effects without being accused of fostering a "science fiction" mentality, i.e., an unrealistic one. The problem, then, is twofold: ignorance of unintended consequences, and ignorance of the *range* of possible outcomes of the unknown unintended consequences. It seems reasonable to anticipate more fruitful discussion about possible changes in the way we work and live if we put *all* our worries on the table, not just a select few that conceal deeper worries. The heart of the issue surrounding most debates on technological change is not the immediate effects of the new "technologies," but rather, how we see these innovations threatening our way of life, i.e., our values.

Given this situation, it seems perfectly reasonable to resist technological change. That is, the justification for the social critics' cautions with respect to the development and introduction of new technologies is to be found in the degree of uncertainty of the impact of this new stuff on our values. And while I think it is perfectly reasonable to resist change for this reason, some cautionary notes for the social critic must be raised as well.

First, as mentioned earlier, ignorance of the consequences by itself is not enough to argue for total opposition to technological change. It is, however, good grounds for asking that more attention be paid to the problem of investigating the possible types of outcomes. The distinction between total opposition and fully considering (to the extent possible) the range of outcomes is crucial. General opposition to technological change is the mark of Luddites. In a world characterized by advances in the tools, systems, and gadgets we call technologies, to oppose new items of these sorts is both unreasonable and dysfunctional. In learning to live in a world where the rate of technological change, if not actually increasing, at least *appears* to be increasing, we must learn to live with it in such a way as to protect ourselves from ourselves. To ask for more information about possible consequences is an important step in that direction.

The kinds of information we will need in order to make the best decisions fall into three general categories: (a) facts, (b) values, and (c) methods. Let us look at each of them separately.

Facts are the most obvious kind of information needed in decision-making situations. But as simple as this sounds, under analysis it becomes a very complicated issue, primarily because there are no facts *simpliciter*. There are a variety of different viewpoints concerning the nature of facts (Fleck 1935; Cohen and Schnelle 1986). Some conceive of facts as being in the world. For others, facts are statements that are made true or false by virtue of correctly asserting what is the case in the world.[5] But what state of the world, and why this fact rather than that turn out to be factors of (b) and (c), as well as theory (see Chapter 1).

What facts are relevant to a decision is hard to sort out. Values play a role here in two different ways. There are epistemic factors at work when we ask such questions as how well established this fact is. To a certain degree, this will be a function of (c), the methods we use to develop, generate, find, categorize, test, etc., the facts. There are also what I call aesthetic factors at work here—values that reflect morals, lifestyle, religion, etc. Thus it is possible to imagine a situation wherein you choose the facts that lend credibility to some future state of affairs you deem desirable.

Facts and values also require theory in a decision-making context. Theory serves to provide criteria of relevance for what is going to count for this situation. The problem here is that, depending on your choice of theory, the "same" facts can play different roles or be evaluated in different ways. We see this happen most often when scientists are called as expert witnesses on different sides of a case. Their testimony can differ precisely because they employ different theories and facts. The situation is

---

5. It is also a topic of some discussion whether there are atomic and general facts, or just one or the other. These issues, however, do not concern us here.

made more difficult because there is no obvious set of criteria for theory choice.[6]

As we can see, the call for more information is not, by itself, a clear-cut panacea. And the situation can become even more complicated. For the most important factor to be considered in the search for the best information on which to base a decision is the method used to generate that information. Many disagreements *appear* to be over "the facts," when the real issue is how those facts were created, generated, found, etc. In short, prior to arguing over the possible consequences of introducing a new technology, we need to investigate the reliability of the methods we use to generate the facts that form the basis for that argument. Once again we find yet another area in which epistemological factors must be settled before social criticism can make serious headway.

Second, the problem of induction notwithstanding, there is no justification for claiming that one set of values is superior to another. That is, there is no demonstrably preferred form of human living. To say otherwise is merely to assert that this set of values is superior because of yet some other set of values. Thus, murder is wrong because I value human life; I value human life because it is the only evidence of intelligence in this universe and I value intelligence; and so on. In short, an argument for the superiority of one set of values over another results either in an infinite regress or in an appeal to authority, both of which are unacceptable. That being the case, there can be no completely justified argument for or against any particular lifestyle. So to resist technological change on the grounds that changing the way we work will change the way we live and this is necessarily bad is itself unjustified.

But from the fact that no absolute justification for any particular set of values is forthcoming, it doesn't follow that we can't arrive at some sort of consensus about some basic values, and maybe even some derivative ones. I take it that this is the drama we see being played out as the Supreme Court works continuously to interpret and reinterpret the U.S. Constitution. The Constitution lays out some basic values that the citizens of the United States have endorsed in some way or another. The Supreme Court interprets those values in the context of a constantly changing social environment in such a way as to maintain consensus. Agreeing on the basics and maintaining that agreement are not easy tasks. But it can be done. What is important here is that we not confuse (a) the fact that we in the United States appear to have an agreement on the fundamentals and a systematic means for revising what it is that we agree to agree on, however

6. To illustrate the confusion over criteria for theory choice, see the various accounts offered by Carnap 1956; Feyerabend 1965; Kuhn 1962; Lakatos 1978; and Laudan 1977.

fragile that agreement may be,[7] with (b) there being a valid argument for the universality of those values. Those are the values on which we have consensus, more or less, nothing more.

The caution to the social critic, then, is this: if you seek to protect those values you express in the face of technological change, remember that no matter how hard you try to predict all the consequences of your actions, past, present, and future, you can't know everything. You must act in a world of uncertainty, and holding out for the *status quo* is no guarantee that everything will turn out right. Second, recognize that not everyone will accept your values and that others are equally well justified in rejecting your claims of superiority. You will have to work toward building a consensus, and this is fundamentally a political activity, not necessarily one governed by reason. And I will go one step further: technological change is a fact of contemporary life. We live in a world where we are continually seeking new ways to improve how we work and consequently how we live. But the challenges this world produces are no different from some imaginary world where there are no technologies, for even in that world we would need to predict the effects of our actions on the future, and we can't do that perfectly.

Finally, something needs to be said about what technological change can mean positively for human development. The whole point of changing the way we work is to open up the possibility of improving our current situation. And this can be understood in a narrow or broad sense of improve. Some discrete change may make it easier to accomplish what is otherwise a boring and unfulfilling task. The broad sense has to do with what technological change ultimately means for humanity. Without getting on a soapbox, I wish to suggest that the more we come to use innovative and labor-saving support systems to improve our work situations, the more we free up our time and energy to explore new and promising avenues of human development.[8] To put it romantically: technological

---

7. Another way to put this is that what we have managed is to agree on the means for arbitrating the impact of various changes. This does not entail agreement on the content.

8. On a personal note: a number of years ago we were fortunate to have Langdon Winner on our campus for a visit. This was just after the publication of his *Autonomous Technology*. In response to his worries about the disenfranchisement of individuals by large corporations, in particular the electric power industry, I suggested the following: speaking as someone who lives on a farm with large forest reserves, and as one who heats his home with wood, consider the time and energy it takes an individual to duplicate the services provided by a power company. First, there is the financial investment: chainsaw (oops—I use gasoline!), pickup truck, wood stove, storage shed, various devices for splitting wood and hauling it from shed to house; time invested in all the above, plus

change makes human growth possible. And while I admit that the prospect of a constantly changing understanding of human nature, yielding no fixed content to the notion of "humanity" other that what we are becoming, can be a scary one, it is also what makes humanity exciting.

---

time required to maintain a heated wood stove; then add the time it takes to go to the woods, chop down trees, dispose of unusable limbs, etc., cart wood home, stack it, etc. Now with electric or gas heat provided by some company, I could read a novel, listen to an opera, groom my dogs—need I go on? Yes, there are certain abstract dangers to aspects of human freedom posed by large electric companies. But what they provide affords the potential for an expansion of other human activities that would not be possible otherwise, including the time to think about the possible dangers large-scale operations such as electric companies pose.

CHAPTER EIGHT

Scientific Change and the Technological Infrastructure of Science

Section 1. Theories of Scientific Change

There are numerous theories of scientific change in the literature. And they all have one thing in common. In accounting for the succession of scientific theories over time, they ignore the contribution of the support systems in which the activities of scientists are embedded. This support system is what in Chapter 7 I called *the technological infrastructure of science.* By the technological infrastructure of science, I mean: *the historically defined set of mutually supporting sets of artifacts and structures without which the development and refinement of scientific knowledge is not possible.* As we saw in Chapter 6, Galileo's development and use of the telescope took place in the context of his role as, first, a professor at Padua and an employee of the Doge of Venice, and then as Chief Mathematician and Philosopher for Cosimo d'Medici of Florence. First the university and then the d'Medici court provided support systems for Galileo to develop the instruments that he used to make the observations that forced significant changes in the theories of the heavens.

What I intend by the use of the phrase "technological infrastructure" is to recognize the social, political, economic, technical, and scientific contexts in which specific scientists are embedded and the contribution these contexts make to their work. Science does not develop and make what progress it does outside of society. By including the social structures that support science in this account, I am merely following through on the argument of Chapter 1, in which I claimed that social structures such as legal systems are as much tools as are hammers and automobiles. But I also intend the phrase "the technological infrastructure of science" to suggest more than merely a recognition of the social dimension of science. With this idea I want to emphasize the extent to which scientific development, which depends so much on new discoveries, is likewise, and for that very reason, dependent on the development of new forms of instrumentation, data processing, and analysis.

Now, in so doing, it is important to acknowledge the pioneer work of the historians and sociologists of science and technology in alerting us to this important aspect of science. In addition, these scholars did more than alert us, they produced a wonderful and fascinating challenge, given their empirical studies (data); what role is left for philosophical theories of science and technology?

According to the sociological critiques,[1] science and technology are both social processes (with which I do not disagree) whose results can be completely explained as the result of negotiation among their participants and other exclusively social processes (with which I disagree). The importance of the work of these social scientists cannot be underestimated. The detailed studies expose, without question, the social dimensions of scientific inquiry. What is in question are some of the conclusions drawn from these studies. In particular, there is one conclusion that seems to bring the credibility of the sociological account into doubt: the claim that any account must "seek to explain the content of scientific knowledge as far as possible in social terms. *Rationality* [whatever that means] must play little part in explaining how the world comes to appear as it does" (Collins 1983, 272). That is, some of these analyses, particularly those of the Strong Programme, deny any role for the physical world in the generation of scientific knowledge. The viability of this form of social realism will be reconsidered later.

On the topic of technological change as a counterpart concept to scientific change, my underlying theme is this: following Derek Price, it seems clear that *progress in science is a direct function of increasing sophistication not merely in instrumentation, but in the technological infrastructure that underlies and makes mature science possible.*

Price claimed: "historically, the arrow of causality is largely from the technology to the science" (Price 1963), but this is only part of the story. By emphasizing the causal priority of technology in scientific progress, Price was attempting to overcome a popular characterization of the relation between science and technology in which technology is placed in a second-class position as the offshoot of science or sometimes its "handmaiden." Price was on the right track, pointing out that despite the fact that historians and philosophers of science from Kuhn (1962) to Laudan (1977) have a tendency to talk about progress in science in terms of the history of ideas, a significant role is played by technology, a role largely

1. Bijker 1987; Bloor 1991; Latour 1987; Pickering 1984; Pinch and Bijker 1987, among others.

ignored by these same philosophers and historians of science. The typical history-of-ideas story of science proceeds by relating that, for example, Newton's mechanics replaced Aristotle's, and then relativity theory replaced Newtonian mechanics.

This story is usually told in Kuhnian fashion, without any mention of the means by which anomalies were discovered. It is merely announced that following a certain experiment, it was decided that so-and-so's theory was false, and it was replaced by another. Thus, a typical bad history would tell you that Michelson's and Morley's experiment was developed to test for aether drift, as predicted by Newton's theory. Once it was discovered that drift did not occur, Newton had to be abandoned. Enter Einstein, and all is saved. Very few histories reveal that Newton did not talk about aether drift; the notion evolved over a hundred years in the course of his successors' efforts to adjust his theory in light of their experience with it. Likewise, very few accounts tell you about the details of the Michelson-Morley experiment.[2] The point here is that on the history of ideas account the history of progress in science is made to read like merely the replacement of one bad theory by another once the bad theory is somehow discovered to be faulty.[3] All of this rests on the presumption that the models of the logic of confirmation produced by philosophers have some bearing on what scientists do, which isn't at all clear. What is ignored is the role in all of this of *the technological infrastructure* within which the falsification and/or confirmation of theories takes place, to the extent that theories are falsified and/or confirmed. A classic example of how the technological infrastructure of science is ignored, and how the progress of science is presented as merely the ideas embedded in theories, can be found in an issue of *Mosaic*, an official NSF publication:

> Every so often, in the long course of scientific progress, a new set of ideas appears, illuminating and redefining what has gone before like a flare bursting over a darkened landscape. It happened when Galileo realized that physical laws needed to be written with numbers and invented the scientific method, when Darwin found an entirely different way to consider the evolution of living things, when Freud placed consciousness and emotion in a new context, when Einstein found a radical way to look at space

2. Or that both experimenters were Americans and that the experiment was carried out in Cleveland, Ohio, at what was then the Case Institute of Technology. After all, with names like "Michelson" and "Morley" they just had to be British and the experiment must have taken place at the Cavendish; didn't they all?

3. This is the case even in philosophical accounts of scientific progress such as Lakatos 1978, and Laudan 1977.

> and time, and when Wegener launched an earth science based on continental drift. (Fisher 1991, 3)

More to the point, prior to the appearance of what Ackermann has called the New Experimentalists (c.f. Ackermann 1985, 1989), philosophers rarely talked about the epistemology of experimentation, or the nature of the link between experiments and the theories they are supposed to test, or the impact of experiment design and the availability of materials, techniques, and instruments, the major exception being Gason Bachelard.[4] This aspect of the story of the progress of science/technology is important particularly at times of dramatic changes, such as are marked by the replacements of one major theory by another, because it is precisely at this juncture that what counts as evidence and how it comes to count as evidence are often at issue.[5]

Section 2. The Technological Infrastructure of Science

I now turn explicitly to the role of the technological infrastructure of science in the growth of knowledge in general. I start by reviewing some features of the manner in which Galileo's development and use of the telescope helped create an initial technological infrastructure for astronomy, and then move to a sketchy reconsideration of that notion as it occurs in modern guise. In so doing, I hope to make plain what is meant by a technological infrastructure of science. Instead of attempting to argue one side or another of the old science/technology debate, I have recast some of the issues so as to demonstrate the epistemological importance of a technological infrastructure construed as interrelated sets of artifacts and structures. Furthermore, just as it makes no sense to talk broadly of technology, it makes no sense to speak of the history and development or importance of a single artifact, suggesting, as this does, that once invented, artifacts remain stable over time. My general thesis is direct: *the development of new information in a mature science is, by and large, a function of its technological infrastructure.* In short, scientific discovery today depends almost completely on the technological context without which modern science would be impossible. I will not raise the question of the merits of this situation until the end of my discussion, although I will provide a hint: in this age of increasingly theoretical science, the technology behind the science may be our only contact with reality, and even so, it is

---

4. Bachelard 1934. This situation is changing. See Franklin 1986; Ackermann 1985; Hacking 1983; Galison 1987, 1997; Cartwright 1989.

5. I have discussed some of these issues elsewhere. See Pitt 1991, chap. 5.

at best a tenuous one. But now let us turn to the relationship between new and improved artifacts and science.

Scientific progress is marked by scientists providing us with better and more coherent views of the nature of the universe and of our role in it. Scientific progress then requires new discoveries. Strangely, in the talk of scientific progress there has been little discussion of discovery by philosophers.[6] To the extent that the issue has been raised, it comes in three contexts. First, there are the problems the concept of discovery creates or sets for cognitive science. I will not discuss these at all. Second, discovery is a problem for realism where the debate hovers over the distinction between discovery and invention. Consider the question, for example, of whether it is possible to discover something that does not materially exist, that is, an idea or a theory. The problem of figuring out what this means rapidly becomes tangled, despite the fact that things seem fairly easy at first. One cannot invent, for example, the Americas—they are already there, so we discover what is there. But scientific theories are invented, not discovered, and yet they are supposed to be about what is there. To say we discover a theory makes it sound as though the theory has been lying around waiting for us—but that is too Platonic for my tastes, especially since we keep "discovering" the wrong theories—i.e., false ones that get rejected eventually. On the other hand, we need to avoid making it sound as though we invent theories out of thin air—surely scientific theorizing has some relation to what is there. Thus there is a certain tension surrounding discussions of discovery in accounts of the development of scientific theories, which tension is generally resolved by invoking a temporal ploy—we *begin* by inventing ways of speaking about situations that have avoided our efforts to understand them until we have some sort of acceptable proof to the effect that what we invented to explain the situation really is there. At that point we say we have discovered these new phenomena, such as gravity or quarks. But this is not really a solution; it is more like a wiggle. I will return to this problem in discussing social constructivism.

The third situation, in which discovery has been a topic for philosophers of science, is as unhappy as the second. It is to be found in the context of Reichenbach's distinction between the context of discovery and the context of justification—a distinction employed so well by Popper in *The Logic of Scientific Discovery.* Popper made things very difficult with his classic dismissal of discovery as an issue for philosophers by characterizing it as a topic fit only for psychology. His own view is frustratingly obscured through the mistranslation of the original German title, *Logik*

6. There are, of course, exceptions. In this case one wants to turn to the work of Thomas Nickles. See Nickles 1985.

der Forschung, as *The Logic of Scientific Discovery* when Popper rejects the very concept of a logic of discovery in the first five pages. Surely we would all have been served better if the title of Popper's book had been translated more accurately as *The Logic of Scientific Research*—for it was the structure of *that* process—*Forschung* in German, meaning "research" or "investigation"—with which Popper was really concerned. But, the follies of mistranslation aside, it is nevertheless true that, for the most part, philosophers of science in the middle years of this century endorsed the Reichenbach/Popper view that discovery is not susceptible to logical analysis and, hence, is not an appropriate topic for any serious discussion of scientific change. It was only some years later, following the publication of Kuhn's *Structure of Scientific Revolutions* when the locus of philosophical attention shifted to the historical process of science and away from the logical positivists' concerns over its rational reconstruction, that discovery once again became an acceptable topic. Only now it posed problems of the second sort noted earlier, i.e., how does a scientific realist deal with the discovery/invention of theoretical entities?

This issue has currently taken on a slightly different shape from the one it had forty years ago. This is not an unexpected turn of events, given that many old problems never really die; they often reappear in a different vocabulary and context, wearing new clothes as it were (c.f. Rorty 1968, 39). Today the invention/discovery battle is taking place between philosophers of science who are scientific realists and sociologists of science belonging to what is euphemistically known as the Strong Programme. Scientific realists believe some version or other of the claim that the theoretical entities mentioned by our best scientific theories actually do exist. Thus, for scientific realists we eventually do discover the real world. There are varieties of realism but they do not concern us now. The practitioners of the Strong Programme, on the other hand, could be said to be inventionists, although they prefer the term "social constructivist." On their view, what most of us call the real world, indicating by that that feature of reality which is independent of us, is nothing more than the result of negotiation among scientists with special axes to grind. For the constructivists, reality is invented or constructed by us; it is not a phenomenon independent of human beings. There can be no question but that there is a social dimension to the activity of science. On this the constructivists are quite correct. But, it is not the case that our account of scientific inquiry can be exhausted by the appeal to the social, for to do so requires that we deny that science is about something, i.e., the physical world. Each of the sciences has a domain within the physical world which its theories and procedures are concerned to investigate. It is the *stuff* of that domain that the investigators are determined to uncover and understand, and it is the stubbornness of that stuff that forces the abandoning of one theory after another as we try to figure out the characteristics and principles governing the behavior of the stuff. To deny to the process of scientific inquiry the very world against which it butts its inquisitive head is to turn scientific inquiry into afternoon tea. Finally, to insist on the absoluteness of the social is to assert a form of social realism, which of course is contrary to the idea that there is no reality to the matter of science. Consider the following dilemma:

(1) If all is social, then there is no world in which the social operates—which is clearly unacceptable; or

(2) if all is social, there is no agreed-upon reality, since for every proposal there is an alternative, hence there is no social domain—equally unacceptable.

Thus, if all is social, either there is no world in which the social can exist, or there is no agreement on what constitutes the social, hence vitiating appeal to the social.[6] Dilemmas are fun because they force the issue, as does this one. In the light of the dilemma one is tempted to avoid any appeal to the social at all. But that is to throw the baby out with the proverbial bath water. We need to find a way to understand the social dimensions of science without turning the social dimensions into a joke. This is where I think the notion of the technological infrastructure of science can be of assistance.

Returning then to the problem of articulating the technological infrastructure of science, we are going to need some definitions. Three, in particular, are relevant to our concerns.

Discovery: the cognitive apprehension of that which has not been so apprehended or apprehended in that manner before.

Technology: humanity at work.

The technology of discovery: humanity at work cognitively apprehending that which has not been so apprehended or apprehended in that manner before.

These definitions present us with a few problems when viewed from the perspective of the realist/constructivist debate. For example, do we, in cognitively apprehending electrons using an electron microscope for the first time, invent or discover electrons? The way to avoid getting stuck

7. Recently James Collier has developed this argument at length. See Collier 1998, chap. 4.

back in the very situation we are trying to avoid is to take our definitions seriously. The definition offered for discovery makes no ontological claims, only an epistemological one. One must "cognitively apprehend something new or in a new way." It doesn't follow that such an act entails that what is cognitively apprehended must exist. Thus this account of discovery, it is to be hoped, avoids the old problems of the realist and the constructivist, at least in the manner in which these were plagued by them. [8]

Turning back to the definitions, I want to lay them out so as to help clarify some of the issues that are before us. But we seem to have both too much and too little in these definitions to be able to understand the role of the technological infrastructure of science. Attending to "cognitively apprehending people at work in a new way" is not going to help us explore the sense in which sets of artifacts generate new scientific discoveries.[9] We need something else; we need to know the manner in which further scientific work depends on new developments in the artifacts, i.e., an account of the *invention* and modification of the relevant artifacts in these circumstances. That is considerably more complicated. It requires our account of "technological infrastructure," characterized as before:

> *A technological infrastructure:* an historically determined set of mutually supporting artifacts and structures that enable human activity and provide the means for its development.

The notion of *mutually supporting* sets of artifacts is difficult to nail down in the abstract. What is ultimately perhaps most important is not the notion that science works within a framework of interrelated sets of artifacts, but the realization, nay, *discovery* that the technological infrastructure has itself grown and developed over time in conjunction with those

---

8. Only if your definition of *knowledge* entails existence would you be back in the old ditch in a hurry. Definitions of knowledge that entail the existence of the things that are known usually invoke a truth condition—such as in "knowledge = justified true belief." Luckily, there exist accounts of knowledge that avoid the problems truth conditions present. For example on my account, (see ch. 1) which I will not belabor here, I distinguish between what is proposed by individuals as candidates for knowledge and the endorsement of those claims by the appropriate social community. An individual may think he or she has found the truth about a particular matter, but thinking or wishing doesn't make it so. Only when the claim has been endorsed by a particular community does it count as knowledge. The criteria the community invokes may have nothing to do with truth—it may, for example, remain satisfied with coherence or with practical efficiency. But, and this is what counts here, if the community determines knowledge, then inevitably truth will go by the board (Pitt 1983). This is the germ that the social constructivist and most relativists exploit.

9. But it may bear on the resolution of discipline-specific problems by importing techniques and individuals from other disciplines.

features of the activity we call science. Thus I am not claiming that science, whenever and however it is or was practiced, has this kind of technological infrastructure. I will argue, however, that the development of a technological infrastructure is essential if science is going to *continue* to provide us with new discoveries about how the universe works. In short, after slow and modest beginnings, a *developed* science *requires* this kind of technological framework. The sorts of investigations and explanations it is called upon to produce require more than mere unaided human thinking alone can produce. I will return to consider the consequences of this claim later. For now, this is enough by way of speculation; let's start to build the case by looking again at Galileo and his telescope.

In Chapter 6 we left Galileo after he had developed a more powerful telescope to confirm his observations of the moon, publishing those observations in *The Starry Messenger*, and in a certain sense sealing the fate of modern astronomy.

In defense briefly of this rather dramatic claim, it should be noted that I am not claiming that Galileo was the first to use an instrument to investigate the heavens and that he thereby forced astronomy onto its historical path. The use of instruments to assist us in our exploration of the heavens already had a long and rich history by Galileo's time. The astrolabe, for example, a device for determining the positions of the planets and the stars, was prominent in Arabic as well as in European astronomy. The quadrant was also a well-used device for determining positions in the heavens.

But unlike the astrolabe or the quadrant, the telescope produced fundamentally new kinds of information. The telescope did not, as did the astrolabe, merely assist in the refinement of measurements according to an established theory. It produced fundamentally new information about the structure and population of the heavens. This information conflicted with the then-established astronomical theories that determined what could and could not be the case astronomically. In this sense then, the telescope forced a transformation in cosmology. The instrument, in effect, required a major overhaul of theory. What was being demanded of theory then in turn forced a reworking and refinement of the instrument, demanding better and more accurate observations to confirm the new theory, which in turn pushed the matter even harder toward further theory revision. A basic new feature had been added to the activity of science: the *interplay* between instruments and theory. It was no longer the case that theories merely used instruments to confirm assumptions. Now we see instruments that are not necessarily coupled to any particular theory (remember that Galileo did not develop the telescope for astronomical purposes), yielding information that forces the scientists defending certain theories to rework their theories in order to accommodate the new data. But, as in any such successful reworking of a theory, the new theory will not only accommodate the new data, it will also suggest opportunities for new discoveries. This then forces revisions of or changes to or even the invention of new instruments, which in turn force the theories to be reworked, etc.

Thus, in Galileo's case, following the first use of the telescope to look at the moon with Cosimo d'Medici, the simple single instrument was to become a complex of instruments. Galileo originally intended his telescope to be handheld for maritime use. But for astronomical purposes it needed a base, then a fixed position from which the observations could be regularized. Tables of sightings could now be corrected, and the need for further refinements in the tables would force refinements in the telescope itself.

For example, a major problem in astronomy was determining the size of the planets. For this, Galileo's telescope, with its concave lens, was not the optimal instrument. In the 1630s it began to be replaced by what van Helden (1989, 113) calls "the astronomical telescope," which had a convex ocular and produced greatly improved clarity in its images. It also had a broader field of vision, which permitted the introduction of a micrometer into the instrument itself, thereby improving the precision of measurements. This was the kind of instrument Huygens used to measure the diameters of the planets. Slowly, Galileo's simple device was becoming a set of things, each part of which could be separately refined, and which in so doing would ramify its effects on the others—perhaps not all the others all the time, but a kind of domino effect was evident. Furthermore, the availability of increasingly precise measurements of particular features of the observable universe also forced changes in the manner in that the relative distances of the planets were calculated. So now we have the instruments and their refinements forcing changes not only in cosmology but in the auxiliary methods that augment it. In this manner, the discovery of the size and structure of the solar system and then the universe was undertaken.

The story could be told without mentioning the instruments. For example, we could say:

> Galileo showed there was more than one center around which planets revolved, forcing a revision of the geocentric theory of the universe. His methods were developed in such a way as to allow for the determination of the distances between planets and the relative sizes of the planets. Modern astronomy continues his program of empirical investigation of the universe.

That says what we have been saying, but the picture it provides of science is, to say the least, impoverished. The mechanism behind the changing ideas is lost, without which we truly have no explanation.

The need for a refined explanation is, I maintain, the proper motivation for including the technological infrastructure of science in our history of culture. For if we want an explanation for the development of science, we need to offer more than a recitation of the sequence of ideas produced by scientists. We need an account of how those ideas were developed and why they were abandoned and/or refined. We are thus dealing with an issue in historiography. An historically sensitive *explanation* of scientific progress and discovery requires appeal to some mechanism. It will not suffice merely to provide a list of dates of events for the selection of those events, for there is a need to justify the selection of those events. Traditionally one selected those events that can be shown to be in a direct line with current developments. I am urging that the events in question be selected for the importance in their own setting, whether or not they eventually lead to something that we deem important today. That is why the traditional history-of-ideas approach, with its base in the present, is inadequate. I am proposing that *the mechanism* that makes the discoveries of science possible and scientific change mandatory *is the technological infrastructure* within which that science operates, and that to understand why a science worked the way it did and why it works the way it does, you need to understand its context, which happens to include in important ways its technological infrastructure. In short, you can no longer do philosophy of science, history of science, or even sociology of science without the philosophy and history of technology.[10]

Perhaps one more look will help us make the case more convincing. Optical astronomy has come a long way since Galileo's little eight-power handheld telescope. We don't need to turn to the Hubble space-based telescope to see that. Not only have telescopes grown in size, but the necessary support systems have become more complicated. The truly large telescopes require massive housings, highly sophisticated background technologies to produce the machines and lenses, electricity to run the equipment and, once cameras were introduced, all the apparatus needed for quality nighttime photography, and the optical theories to support interpretations of the products, computers to calculate position, manage the photography and coordinate the systems, and the list goes on.

Astronomy is the science of the heavens. Its function is to describe the universe in terms of the relative constitution of and positions of its parts. To accomplish this goal, astronomers need to be able to see the heavens. And so we have the elaborate technological infrastructure of the optical telescope. But to assume that the components of the universe are limited

10. It is ironic that in their search for a universal model of scientific explanation, philosophies of science have insisted on capturing causality, and yet, when it comes to explaining scientific change they have not. Instead, the emphasis has been on the rationality of selection procedures, with the assumption that rationality has caused efficacy of rationality is hardly supported by history.

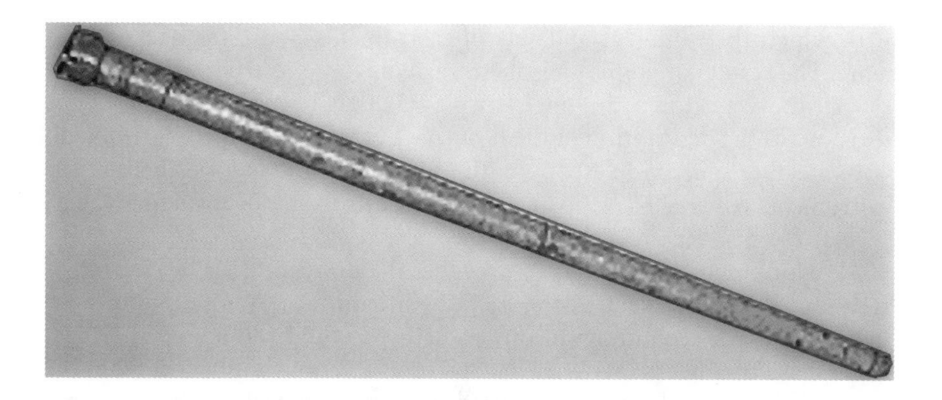

to those that can be seen by the human eye is absurdly homocentric. So if you add to the optical infrastructure radio telescopes, and the theories upon which they are based, spectral telescopes, the use of high-speed computers not only to control the telescopes, but to generate and interpret at least the first- and second-order information they generate, the computers and the computer programs necessary for all that, the launching of space-based telescopes and the technological systems behind that, the infrastructure behind the computers, etc., and more—if you add all that in, the technological infrastructure of astronomy appears to swamp the science. But there is still more, for at each stage, the development of the instruments is constrained by the fit with other instruments and the theories with which they interact and sets of instruments and their backup systems. The results of employing these systems force restructuring of theories all the way down the line. It isn't just that new observations force revisions in the description of the heavens. The questions include: How do you integrate spectral telescopy with optical? Do the theories behind the instruments cohere? How do anomalous results from one instrument, e.g., excessive red shift, affect the other theories?

We look with awe at the pictures that the new space probe, appropriately called Galileo, sent of the Earth while on its way to Jupiter. If we think about the technological infrastructure behind the pictures, we get some sense of what is involved. The pictures are not simply sent from the space vehicle, traveling at high speed in its own trajectory, but they are sent to Earth, which is also traveling at high speed and on its own different trajectory. The "pictures" are transmitted as electronic code. That means they have to be disassembled, sent, reassembled, etc. The machinery, the programming, and the capacity for mistakes are enormous. If you add the testing of scientific theories to the problem, and the interaction between the theories and the technological infrastructure as well as among themselves, there can never again be a simple history of the ideas of science, nor should there be.

If the science is astronomy, or even cosmology, then we must turn to the technological infrastructure to understand its results. It is no longer possible to say "science tells us . . . ," and it is certainly misleading to say "science and technology tell us . . ." for no one has taken the time to spell out what that means. When we do spell it out, we will find that what we really wanted to say was: "The technological infrastructure within which scientific theories are being developed and transformed makes it possible for us to describe and explain the universe in the following way." This contextualization of our science is extremely important. The kinds of things we come to know about the universe, or to put it more dramatically, the universe modern technologies reveal to us, are a function of this complex interaction between theories and technological infrastructure. Furthermore, it would seem that with a different technological infrastructure, "science" might yield a different universe—or would it? As we attempt to answer this question, we also find ourselves back in the realism debate. Let us then turn to realism briefly one final time.

To answer this question requires a definition of "real," which is what the many versions of realism purport to provide. Rather than rehearse the various efforts to nail down this elusive notion, I want to point to a systematic problem they all share and then to propose an alternative realist account that both avoids that systematic problem and offers a more intuitively plausible program for understanding scientific and technological change.

Common to all versions of scientific realism is the notion of reduction. That is, whatever else is the case, the bountiful variety of *things* and/or *processes* is to be understood as deriving from the fundamental things or laws that the realism endorses. I do not believe that there is a justification for the methodological appeal to reduction. At best the requirement behind reduction is an appeal to simplicity. But again, there is no justification for the assumption that the universe must be simple. If we assume that there can be a mathematically simple account of the beginnings of the universe, it doesn't follow that the universe as we know it now must abide by principles of simplicity, and if we reject simplicity and reduction we end up with my form of realism: Sicilian realism.[11]

If we take as our starting point the fundamental claim of scientific realism—namely, that theoretical entities are real, however glossed—and couple it with the historical awareness that theories change and are replaced, we have a problem: which theoretical entities from which theories are really real? A Sicilian realist says, with some important caveats, that they all are. Sicilian realism is realism with a vengeance; the universe is a very complicated place, to echo Marjorie Grene and Richard Burian.[12] What we manage to do with one theory/technological infrastructure is to cut the universe at one of its many joints. Optical telescopes tell us planets and stars are real. Radio telescopes tell us there is more out there. Sicilian realism admits all of this. What Sicilian realism does not admit as at all necessary is the kind of reduction that more standard forms of realism assume. Thus, atoms, electrons, and quarks are all equally real without one having to be reduced to another. Seeing the universe in terms of atoms is a function of cutting it only one way, and there are others.

"But surely," runs the objection, "you can't mean that the entities of any and every theory, from the beginnings of theorizing to the present, are real!"[13] Clearly that is correct—I can't possibly mean that. I do not mean to include the theoretical entities of all theories for all time. Rather, we are speaking of only those that stand together to form a coherent whole in their best configuration at a particular historical point in time. Fundamentally, Sicilian realism rejects reductionism in the form represented by the difficulty with Eddington's table. I want to argue that we don't eliminate

11. I use the phrase "Sicilian realism" to reflect the rich and varied cultural history of Sicily. For centuries, Sicily was at the center of trade routes and invasion paths. It was occupied by conquering armies and friendly forces coming from all directions. It was a major launching point from which the crusades left Europe and headed for the Middle East. The composition of Sicilian culture reflects Greek, Roman, German, and Moorish input, among others. In its architecture, music, food, and customs it is a multilayered and complex place, much like our universe.

12. In private conversation.

13. I thank Gary Hardcastle for forcing me to deal with this problem.

the table in favor of a set of electrons, and then if we find even smaller entities, reject the electrons in favor of them. Matter aggregates in different ways depending on the size of the components. To deny the existence of any particular aggregate because it is not the smallest seems capricious. The problem stems from the language of the physicists in their search for what they call the most fundamental particles. The confusion, then, is between (a) the most fundamental and (b) the most real. To see this, all one has to do is to attend to the metaphor that accompanies talk of "most fundamental entities": the building blocks of the universe. If we take the metaphor seriously and follow it through, the problem should disappear. Let us assume for the moment that bricks are the smallest component of a building. We use the bricks to construct the walls that encase the space that, together with the walls and roofs, etc., constitutes the building. To then turn around and declare that there is no building, i.e., that there are only bricks, seems silly. No one will deny that the building is built of bricks, but the building is what we are talking about. Likewise for quarks and tables. This is not to argue against the methodology of physics, which uses appeals to simplicity and idealization to guide its searches. It is merely to caution against confusing methodological heuristics with ontology.

To make the point one more time. Even if we agree on what the fundamental entities of the universe are, we still live in a world of planets and stars. Those fundamental units aggregate in certain ways that will be described by a different theory from the theory in which the behavior of the fundamental particles is described. And the aggregates of the fundamental particles will form yet a different and more complex set of aggregates with their own theory, etc. What Sicilian realism argues for is the reality of each set of aggregates. It is a form of entity realism. And the theories that we endorse are the best theories for discrete aggregate sizes. And the criteria for being the best theory are (1) being confirmed by evidence, and (2) coherence with the other theories of the other aggregates in their best-developed form.

What we have to face is the fact that while there is no one necessary way to investigate nature, the mechanisms—read "technological infrastructures"—we develop to assist us set a complicated process in motion in which imagination and creativity are sparked and fed by the interplay between idea and artifact. Artifacts stimulate us to seek uses for them—how to couple them with other artifacts; they also present us with the problem of interpreting the results. Given different sets of artifacts—by definition different simulations—we get different results. But we start small and go large in quick order. Compare Galileo's simple telescope with the complex that we need for a modern mountain-top observatory.

What are the consequences of accepting this characterization of the role of our technologies? Is it not the case, as I am sure some determinist will be sure to suggest, that it means not only that society is run by technology, but that now science is too! No, that is not the proper conclusion to draw. It is not a question of which disembodied and reified nonentity, science or technology, controls anything. What a careful look at history will show is that *as instruments are made more complex by individuals with specific objectives in mind (objectives sometimes, but not always, generated by endorsing certain theories), a complex of interrelated activities develops through which, by choosing certain ways to augment the technological infrastructure, certain options are opened or shut for theoretical testing and exploration.* People still make the choices, and they may choose badly, taking us down a dead end. Or they may opt for a system that does not have the backup to support it. This is what happened to the nineteenth-century astronomer, William Herschel. He built a forty-inch telescope that was certainly a technological marvel. Only there were severe problems. The mounting for it proved unstable. The mirror was made of metal and lost its reflective capacity. It fell into disuse.

My point is that if you want to explain the changing claims and face of science, you have to go beneath the ideas to the technological infrastructure and then you have to unravel the interactions between its parts and the mass of theories with which it is involved. It is that complex that makes it possible to apprehend new things or to apprehend things previously known, but in a new way. The discovery of structures in nature is a function of this complicated, mutually interacting set of artifacts, ideas, systems, and, of course, men and women. Telling that story puts us in a position to understand finally the nature of the scientific enterprise and how it generates new information. It should also alert us to the sensitivity of the system. With so much depending on so much, there are many opportunities for things to go wrong. Likewise, because of the complex of interdependent relationships, the determination of the accuracy or even of the import of the new information this system generates is not an easy task. Scientific knowledge becomes more tenuous and more dearly bought as the technological infrastructure grows, but it is increasingly impossible without it. Sometimes all we know is that this or that system works; we may not know what it is telling us. If I am right, this account also has serious ramifications for our account of knowledge. It is clear that in this complex of interacting systems of supports, the individual becomes increasingly less capable of being the sole generator of knowledge. Increasingly we need to look to the processes by which groups of scientists and engineers achieve consensus on criteria, methods, and such important details as the calibration of instruments.[14] And as we do so, we find the antidote to the disease of despair caused by the recognition that, due to the

14. See Franklin 1997.

manner in which the production of scientific knowledge is embedded in a technological infrastructure, our knowledge of the world is the result of layers of complicated epistemic restrictions resulting in a profound epistemic distance between our knowledge and the world that is its object. Once we realize that it is groups of scientists and engineers that establish the parameters and criteria for knowledge, the feeling of epistemic impotence ought to recede. If we as philosophers remove the burden of producing knowledge from individuals and place it in the hands of groups (where it already lies) of investigators, we increase the probability that viable options will be explored and procedures scrutinized. By insisting on group consensus, we in fact become alert to the complications that the technological infrastructure creates, and we increase the probability that we will become inevitably constrained by the structure of the system.

Section 3. Conclusion

I have looked at technological change as a counterpart concept to scientific change. I have argued that understanding scientific change requires putting the science in context, and one of the more important contexts—for there are many—in which to see science, especially mature science, is its technological infrastructure. The strong conclusion emerging from this discussion is that there can be no analysis of changes in a well-developed science without considering a massive support system of artifacts and social institutions. Thus the growth of science can be seen in similar terms as the growth of human culture, that is, made possible by the tools and mutually interactive support systems we have come to call technology. By shifting the burden of the production of knowledge from individuals to groups, we also take steps toward insuring that the infrastructures do not constrain the search for knowledge but, in fact, help create the dynamics that makes the relations between the sciences and our technologies so exciting. So what is the relation between science and technology? There isn't one; there are many, and that is the way it ought to be.

Bibliography

Ackermann, Robert (1985). *Data, Instruments, and Theory.* Princeton: Princeton University Press.

Ackermann, R. (1989). "The New Experimentalism." *British Journal for the Philosophy of Science* 40: 185-90.

Adams, Douglas (1983). *Life, the Universe, and Everything*. New York: Harmony Books.

Anonymous (1983). "Bridge Disaster Is Tip of Iceberg." *U.S. News & World Report* 95: (July 11, 1983): 8.

Bachelard, Gaston (1934). *The New Scientific Spirit*. Trans. by Arthur Goldhammer. Boston: Beacon Press.

Bernstein, Richard (1993). *The New Constellation: The Ethical-Political Horizons of Modernity/Postmodernity*. Cambridge, Mass.: MIT Press.

Berreby, David (1992). "The Great Bridge Controversy." *Discover* 13 (February): 26-33.

Bijker, Wiebe (1987). "The Social Construction of Bakelite: Towards a Theory of Invention." In *The Social Construction of Technological Systems: New Directions in the Sociology and History of Technology,* ed. Wiebe Bijker, Thomas Hughes, and Trevor Pinch. Cambridge, Mass.: MIT Press. 159-87.

Bijker, Wiebe, Thomas Hughes, and Trevor Pinch, eds. (1987). *The Social Construction of Technological Systems: New Directions in the Sociology and History of Technology.* Cambridge, Mass.: MIT Press.

Bloor, David (1991). *Knowledge and Social Imagery,* 2d ed. Chicago: University of Chicago Press.

Bonner, John (1995). *Economic Efficiency and Social Justice: The Development of Utilitarian Ideas in Economics from Bentham to Edgeworth*. Brookfield, Vt.: E. Elgar.

Bromberger, Sylvain (1992). *On What We Know We Don't Know: Explanation, Theory, Linguistics, and How Questions Shape Them*. Chicago: University of Chicago Press.

Bucciarelli, Louis (1994). *Designing Engineers*. Cambridge, Mass.: MIT Press.

Carnap, Rudolf (1956). "Empiricism, Semantics, and Ontology." In *Meaning and Necessity*, 2d ed. Chicago: University of Chicago Press. 205-11.

Cartwright, N. (1989). *Nature's Capacities and Their Measurement*. Oxford: Clarendon Press.

Chaisson, Eric J. (1994). *The Hubble Wars*. New York: HarperCollins.

Chernousenko, V.M. (1991). *Chernobyl: Insight from the Inside*. New York: Springer-Verlag.

Clark, A., and P.J.R. Millican, eds. (1996). *The Legacy of Alan Turing, Volume 2: Connectionism, Concepts and Folk Psychology*. New York: Oxford University Press.

Cohen, Robert, and Thomas Schnelle, eds. (1986). *Cognition and Facts: Materials on Ludwik Fleck*. Boston: D. Reidel.

Collier, James (1998). "The Structure of Meta-Scientific Claims: Towards a Philosophy of Science and Technology Studies." Ph.D. Dissertation; Virginia Polytechnic Institute and State University.

Collins, H.M. (1981). "Stages in the Empirical Programme of Relativism." *Social Studies of Science* 11: 3-10.
Collins, H.M. (1983). "The Sociology of Scientific Knowledge: Studies in Contemporary Science." *Annual Review of Sociology* 9: 265-85.
Collins, H.M. (1985). *Changing Order: Replication and Induction in Scientific Practice*. London: Sage.
Collins, H.M. (1990). *Artificial Experts: Social Knowledge and Intelligent Machines*. Cambridge, Mass.: MIT Press.
Constant, E. (1980). *The Origins of the Turbojet Revolution*. Baltimore: John Hopkins University Press.
Derrida, Jacques (1967). "Structure, Sign and Play in the Discourse of the Human Sciences." In *Writing and Difference*. Trans. by Allen Blass. Chicago: University of Chicago. 278-93.
Dewey, John (1929). *The Quest for Certainty*. New York: Minton, Balch.
Diesing, Paul (1991). *How Does Social Science Work? Reflections on Practice*. Pittsburgh: University of Pittsburgh Press.
Drake, Stillman (1978). *Galileo at Work*. Chicago: University of Chicago Press.
Drake, Stillman, trans. (1978). *Galileo Galilei, Operations of the Geometric and Military Compass*. Washington, D.C.: Smithsonian Institution Press.
Dupré, John (1996). "Metaphysical Disorder and Scientific Disunity." In *The Disunity of Science: Boundaries, Contexts, and Power*. Edited by Peter Galison and David Stump. Stanford: Stanford University Press.
Ellul, Jacques (1964). *The Technological Society*. New York: Vintage Books.
Feyerabend, Paul (1965). "Problems of Empiricism." In *Beyond the Edge of Certainty. Pittsburg Symposium.* Edited by Robert Colodny. Englewood Cliffs, N.J.: Prentice-Hall.
Fisher, Arthur (1991). "A New Synthesis Comes with Age." *Mosaic* 22, no. 1: 3.
Fleck, Luwdik (1935). *Genesis and Development of a Scientific Fact*. Edited by T. Treun and R. Merton. Trans. by F. Bradley and T. Treun. Chicago: University of Chicago Press.
Foucault, Michel (1966). *The Order of Things: An Archaeology of the Human Sciences*. New York: Vintage.
Franklin, Allan (1986). *The Neglect of Experiment.* Cambridge, England: Cambridge University Press.
Franklin, Allan (1997). "Calibration" in *Perspectives on Science* 5, no.1: 31-80.
Galilei, Galileo (1610). *The Starry Messenger.* In *Discoveries & Opinions of Galileo*. Trans. by Stillman Drake. New York: Anchor Books.
Galilei, Galileo (1632). *Dialogue Concerning the Two Chief World Systems*. Trans. by Stillman Drake. Berkeley: University of California Press.
Galilei, Galileo (1638). *Two New Sciences*. Trans. by Stillman Drake. Madison: University of Wisconsin Press.
Galison, Peter (1987). *How Experiments End*. Chicago: University of Chicago Press.
Galison, Peter (1997). *Image and Logic: A Material Culture of Microphysics*. Chicago: University of Chicago Press.
Hacking, Ian (1983). *Representing and Intervening*. Cambridge, England: Cambridge University Press.
Haraway, Donna (1991). *Simians, Cyborgs, and Women: The Reinvention of Nature*. New York: Routledge.
Harding, Sandra (1986). *The Science Question in Feminism*. Ithaca, N.Y.: Cornell University Press.
Hausman, Daniel, and Michael McPherson (1996). *Economic Analysis and Moral Theory*. Cambridge, England: Cambridge University Press.
Heidegger, Martin (1954). "The Question Concerning Technology." Trans. by William Lovitt. In *Basic Writings*, revised and expanded edition. Edited by David Krell. New York: HarperCollins.
Hempel, Carl, and Paul Oppenheim (1948). "Studies in the Logic of Explanation." *Philosophy of Science* 15: 135-75.
Kelly, Walt (1985). *Outrageously Pogo*. Edited by Mrs. Walt Kelly and Bill Crouch, Jr. New York: Simon and Schuster, Inc.

Kuhn, Thomas (1962). *The Structure of Scientific Revolutions* 2d ed. Chicago: University of Chicago Press.

Lakatos, Imre (1978). *Methodology of Scientific Research Programs*. Edited by John Worrall and Gregory Currie. Cambridge, England: Cambridge University Press.

Lakatos, I., and A. Musgrave, eds. (1970). *Criticism and Growth of Knowledge*. Cambridge, England: Cambridge University Press.

Latour, Bruno (1987). *Science in Action: How to Follow Scientists and Engineers through Society*. Cambridge, Mass.: Harvard University Press.

Laudan, Larry (1977). *Progress and Its Problems*. Berkeley: University of California Press.

Layton, Edwin (1974). "Technology as Knowledge." *Technology and Culture* 15: 31-41.

Layton, Edwin (1987). "Through the Looking Glass or News from Lake Mirror Image." *Technology and Culture* 28: 594-607.

Lewis, C.I. (1946). *An Analysis of Knowledge and Valuation*. La Salle, Ill.: Open Court.

Longino, Helen (1990). *Science as Social Knowledge: Values and Objectivity in Scientific Inquiry*. Princeton: Princeton University Press.

McDermott, John (1969). "Technology: The Opiate of the Intellectuals." *New York Review of Books*, July 31, 25-35.

McDonald, Duff, Pablo Galarza, and Sarah Rose (1998). "Investing's New Frontier." *Money* 27, no. 9: 82-98.

McMullin, Ernan (1968). *Galileo, Man of Science*. New York: Basic Books.

Merton, Robert K. (1970). *Science, Technology and Society in Seventeenth-Century England*. New York: Harper and Row.

Merton, Robert K. (1973). *The Sociology of Science: Theoretical and Empirical Investigations*. Edited by N. W. Storer. Chicago: University of Chicago Press.

Mesthene, E. (1970). *Technological Change*. New York: Mentor.

Millican, P.J.R., and A. Clark, eds. (1996). *The Legacy of Alan Turing, Volume 1: Machines and Thought*. New York: Oxford University Press.

Mills, C. Wright (1959). *The Sociological Imagination*. New York: Oxford University Press.

Minas, Jay (1973). "Emergent Utilities." In *Science, Decision and Value, Proceedings of the Fifth University of Western Ontario Philosophy Colloquium*. Edited by J. Leach, R. Butts, and G. Pearce. Boston: D. Reidel.

Mumford, L. (1963). *Technics and Civilization*. New York: Harcourt Brace and World.

Newton, Issac (1726). *Principia Mathematica*. Cambridge, Mass.: Harvard University Press.

Nickles, Thomas (1985). "Beyond Divorce: Current Status of the Discovery Debate." *Philosophy of Science* 52: 177-206.

Page, Clint. (1983). "Bridge Collapse Puts Focus on Nation's Infrastructure." *Nation's Cities Weekly* 6 (July 4, 1983): 1.

Paglia, Camille (1990). *Sexual Personae: Art and Decadence from Nefertiti to Emily Dickinson*. New Haven: Yale University Press.

Peirce, Charles Sanders (1934). *Collected Papers of Charles Sanders Peirce*, in 8 volumes. Edited by Charles Hartshorne and Paul Wise. Cambridge, Mass.: Harvard University Press. (Abbreviated: CP:Volume.Paragraph)

Petroski, Henry (1991). "Still Twisting." *American Scientist* 79: 398-401.

Pickering, Andrew (1984). *Constructing Quarks: A Sociological History of Particle Physics*. Chicago: University of Chicago Press.

Pinch, Trevor, and Wiebe Bijker (1987). "The Social Construction of Facts and Artifacts, Or How the Sociology of Science and the Sociology of Technology Might Benefit Each Other." In *The Social Construction of Technological Systems: New Directions in the Sociology and History of Technology.* Edited by Wiebe Bijker, Thomas Hughes, and Trevor Pinch. Cambridge, Mass.: MIT Press. 17-50.

Pitt, J.C. (1978). "Galileo: Causation and the Use of Geometry." In *New Perspectives on Galileo*. Edited by R.E. Butts and J.C. Pitt. Dordrecht: D. Reidel. 181-92.

Pitt, J.C. (1981). *Pictures, Images and Conceptual Schemes*. Dordrecht: D. Reidel. 129.

Pitt, J.C. (1982). "The Role of Inductive Generalizations in Sellars' Theory of Explanation." In *Theory and Decision* 13: 345-56.

Pitt, J.C. (1983). "The Epistemological Engine." *Philosophia* 32: 77-95.

Pitt, J.C. (1986). "The Character of Galilean Evidence." In *PSA*. 1986, pp. 125-134.
Pitt, J.C. (1987). "The Autonomy of Technology." In *Technology and Responsibility*. Edited by Paul T. Durbin. Dordrecht: D. Reidel. 99-114.
Pitt, J.C., ed. (1988). *Theories of Explanation*. New York: Oxford University Press.
Pitt, J.C. (1991). *Galileo, Human Knowledge and The Book of Nature: Method Replaces Metaphysics*. Dordrecht: Kluwer.
Pitt, J.C. (1994). "Philosophical Methodology, Technologies and the Transformation of Knowledge." In *Technology and Ecology.* Edited by Larry A. Hickman and Elizabeth F. Porter. Carbondale, Ill.: Society for Philosophy and Technology. 1-26.
Popper, K. (1963). *The Logic of Scientific Discovery.* London: Hutchinson.
Price, Derek de Solla (1963). *Big Science, Little Science*. New York: Columbia University Press.
Putnam, Hilary (1978). *Meaning and the Moral Science*. London: Routledge and Kegan Paul.
Rorty, R., ed. (1968). *The Linguistic Turn*. Chicago: University of Chicago Press.
Rouse, Joseph (1996). *Engaging Science: How to Understand Its Practices Philosophically*. Ithaca, N.Y.: Cornell University Press.
Rudner, Richard (1953). "The Scientist *qua* Scientist Makes Value Judgments." *Philosophy of Science* 20: 1-6.
Schubert, G. (1965). *Judicial Policy Making*. Glenview, Ill.: Scott, Foresman.
Sellers, W. (1963a). "Philosophy and the Scientific Image of Man." In *Science, Perception, and Reality*. London: Routledge and Kegan Paul.
Sellars, W. (1963b). "Induction as Vindication." *Philosophy of Science* 31: 198-231.
Sellars, W. (1968). *Science, Perception, and Reality*. London: Routledge and Kegan Paul.
Shcherbak, Iurii (1989). *Chernobyl: A Documentary Story*. Translated by Ian Press. New York: St. Martin's.
Shea, William (1972). *Galileo's Intellectual Revolution*. London: Macmillan.
Skinner, B.F. (1971). *Beyond Freedom and Dignity*. New York: Knopf.
Skinner, B.F. (1976). *Walden Two*. New York: Macmillan.
Smith, Robert (1989). *The Space Telescope*. New York: Cambridge University Press.
Stout, Jeffrey (1988). *Ethics after Babel : The Languages of Morals and Their Discontents*. Boston: Beacon Press.
Teich, Albert H., ed. (1997). *Technology and the Future.* 7th ed. New York: St. Martin's.
U. S. Congress (1994). *Hubble Space Telescope: Hearing Before the Subcommittee on Science, Space, and Technology, U. S. House of Representatives, 103rd Congress, First Session, November 16, 1993*. Washington: U. S. Government Printing Office.
Van Helden, Albert (1989). "The Telescope and Cosmic Dimensions." In *Planetary Astronomy from the Renaissance to the Rise of Astrophysics*. Part A. Edited by R. Taton and C. Wilson. Cambridge, Mass.: Cambridge University Press.
Vincenti, Walter (1988). *What Engineers Know and How They Know It: Analytical Studies from Aeronautical History*. Baltimore: John Hopkins University Press.
Waldrop, M. Mitchell (1990). "Hubble Hubris: A Case of 'Certified' Blindness." *Science* 250: 1330.
Wallace, William A. (1992). *Galileo's Logic of Discovery and Proof: The Background Content and Use of His Appropriated Treatises On Aristotle's Posterior Analytics*. Boston: Kluwer Academic Publishers.
Westfall, Richard (1958). *Science and Religion in Seventeenth-Century England*. New Haven: Yale University Press.
Winner, Langdon (1977). *Autonomous Technology: Technics-Out-of-Control as a Theme in Political Thought*. Cambridge, Mass.: MIT Press.
Winner, Langdon (1986). *The Whale and the Reactor*. Chicago: University of Chicago Press.
Wirkmann, Debra. (1991). "The Fall of the Tacoma Narrows Bridge." *Exploratorium Quarterly* 15, Summer: 30-34.

Index

Abortion, 112
Ackermann, 125
Aesthetic factors, 118
Aesthetic values, 83, 103
Aether drift, 124
Archimedean mechanics, 96
Aristotelian biology, 92
Aristotle, 3, 124
    on heavenly sphere, 95
    on theory of causes, 68
Artifacts, 30
    failed, 45
    mutually supporting sets of, 129
    performance measures for, 56
    role of, 136
    working of, 45
Assessment, of ideology, 81–82
Assessment feedback, 14, 15
Astrolabe, 130
Astronomy, technological infrastructure of, 132–34
Autonomy
    concept of, 98
    trivial, 88–90

Bachelard, Gaston, 125
Berkeley, George, 3
Bernstein, Richard, 67
Big Science, 9, 28
Big Technology, 110
    defined, 31
    vs. Little Technology, 30
Blame, assessment of, 53
Bucciarelli, Louis
    on design process, 55–56
    model of, 51, 53
    on social context, 60, 65
Burian, Richard, 135

Calculus, role of, 97
Cambridge, Mass., moratorium in, 109
Candidate-claims, 4
Cause(s), Aristotle's theory of, 68
Challenger disaster, 103
Chernobyl, nuclear disaster at, 45, 46, 74
Circumstances, understanding, 19
Clavius, 94–95
Cognitive values, defined, 83, 103
Colvin, William, 57, 58–64
Common sense, systematic application of, 91–93
Commonsense Principle of Rationality (CPR).
    See CPR

Community
    as determiner of knowledge, 4
    and rationality, 20
Component design, subdivisions of, 37
Compromise(s), 85, 86
    and community decisions, 89
Concepts, logic of, 7–8
Conceptual barriers, artificial, 114
Conceptual framework, 78
    and domains of concern, 79
Conceptual scheme, and ideology,76–78, 79
Confirmation, problem of, 7
Connecticut, collapsed bridge in, 47
Consensus, need for, 119, 120
Constitution, basic values in, 119
Constructivists, relativism of, 5
Context
    and social norms, 18
    and transferability of technological knowledge, 41
Convergent realism, 5
"Cookbook engineering," 38
Copernican mathematical astronomy, 95
Copernicus, 93
Cosimo, Grand Duke, 94
    See also Medici, Cosimo d'
Counterpart technological concepts, 25
    questions re, 26–28
CPR (Commonsense Principle of Rationality),
    22–23, 34, 50
    feedback function of, 64

Decision(s), as first-order transformations,
    13
Decision making
    as ideological process, 85
    in MT model, 60
    role of values in scientific, 82–83
    in technological processes, 56
Deductive-Nomological Theory of Explanation.
    See DN
Deliberate process, structure of, 16
Democracy
    forms of, 104–5, 108
    technology as threat to, 100, 102, 107
    as value-neutral, 107
Descartes, René, 3
Design
    defined, 36
    overall, 37
    as social process, 55–56

Design knowledge
  Vincenti on, 36–37
  See also Engineering knowledge; Knowledge; Scientific knowledge; Technological knowledge
Design process, 31
  breakdown in, 64–65
  movement of, 37,55
  role of, in technology, 54
  stages of, 54
Diablo Canyon nuclear reactor, 72–75, 80
Dialogue on the Two Chief World Systems (Galileo), 97
Discourses on Two New Sciences (Galileo), 97
Discovery
  context of, 126–27
  definition of, 128
  technology of, 128
DN, 42, 45
DNA, recombinant, 109–11
  moral questions re, 113
Doge of Venice, 122
Domains of concern, 79
Drake, 93–94

Eddington's table, 135
Einstein, Albert, 89
Election forecasts, 100
Ellul, Jacques, 87
Empiricism, 3–4
End(s), intrinsic vs. extrinsic, 15
Enframing, Heidegger's use of term, 69
Engineering
  as a practice, 36
  Rogers/Vincenti on, 35–36
Engineering knowledge, 32, 35–38
  as form of technological knowledge, 34
  as task-specific, 38
  See also Design knowledge; Knowledge; Scientific knowledge; Technological knowledge
Engineering *simpliciter*, 31
Entity realism, 136
  See also Realism
Experts
  reliance on, 105–6
  use of, 106
Extrinsic ends, 15

Facts
  method of obtaining, 119
  nature of, 118
First Amendment, 84–85
First-order transformation process, 13, 14
Ford, Henry, 89

Galileo, 130, 131
  support systems of, 122
  and telescope, 5–6, 92, 93–96, 98–99
Galileo (space probe), 134
Galileo at Work (Drake), 93
Geometric method, 97
Geometry
  Galileo's use of, 96, 97
  as technology, 96–97
  as theory-independent, 97
Gestalt switches, 95
Golden Fleece awards, 110–11
Grene, Marjorie, 135
Gross Error Test, 62
Group consensus, 138
  See also Consensus

Hall, A.R., 34
Heidegger, Martin, 67–70
Hempel, Carl, 42
Herschel, William, 137
Hippocratic Oath, 110
Homo economicus, 17
House of Representatives, inquiry of, 58
HST. *See* Hubble Space (Optical) Telescope
Hubble Space (Optical) Telescope (HST), 51–65
Hume, David, 3
  on knowledge, 21
Huygens, Christiaan, 131

Ideology
  vs. conceptual scheme, 76–78, 79
  and point of assessment, 81
  quasi-pathological, 73
  technology and, 70–82
Incommensurability, as problem, 38–40, 41
Information, new, as function of its technological infrastructure, 125–26
Informed vote, 105
In principle concept, 105
Input/output assessment model, 23
  See also MT
Instruments, and theory, 130
Interferogram(s), 62, 63
Interventionists, 127
Intrinsic ends, 15
Invention/discovery battle, 127
Inverse Null Corrector, unexpected results from, 60–61
In vitro fertilization, 112
Irrationality, 21–22

Johns Hopkins University, 58
Judgments, as value judgments, 82
Justice, as end, 15

Knowledge, 16
  acquisition of, 22–23
  community as determiner of, 4
  Hume's approach to, 21
  as justified true belief, 4
  pure/applied, 2
  as secondary to acting, 104
  as tool, 92
  See also Design knowledge; Engineering knowledge; Scientific knowledge; Technological knowledge
Kuhn, Thomas, 8, 38, 123, 127
  paradigm shifts of, 95

Land Grant Colleges, 32
Landucci, Benedetto, 94
Laudan, 123
Layton, Edwin, 34
Leibniz, Gottfried, 3
Lipperhey, Hans, 97
Little Science, 28
Little Technology, vs. Big Technology, 30
Locke, John, 3
Logic, 21
Logical Positivism, 7, 40
Logic of Scientific Discovery, The (Popper), 126–27
Logic of Scientific Research, 127
Logik der Forschung (Popper), 126–27
Long Island nuclear reactor, dismantling of, 101, 107

McDermott, John, 90, 91
Major-component design, 37
Majority rule, 104–5
Manhattan Project, 49
Mathematical astronomies, 95–96
Meaning, theory of, 39
Mechanism, epistemological relations between theory and, 6
Medici, Cosimo d', 122, 131
  See also Cosimo, Grand Duke
Mesthene, Emmanuel, 10–11, 35, 36, 90
Meta-paradigm, 39
Michelson, 8, 124
Michelson-Morley, 8
Morality, technology and, 112–13
Moral values, 83, 103
Morley, 8, 124
Morrill Act, 32
Mosaic, 124–25
MT model, 52
  and decision making, 60
  purpose of, 16
  on technology as process, 93
  See also Input/output assessment model
Myth of the Engineer, 52, 53

NASA (National Space and Aeronautics Administration), 57, 60
Nature, as standing reserve, 70
New Experimentalists, 125
New Science, 3
Newton, Isaac, 124
  calculus of, 97
Nonconformance report, 60
Normal design, 36
  divisions of, 37
Nuclear research, military funding of, 111

"On the Question of Technology" (Heidegger), 68, 69
Oppenheim, Paul, 42

Padua University, 122
Paradigm(s), 8
  rules of, 39
Paradigm shifts, and Gestalt switches, 95
Peirce, Charles Saunders, 4–5
People, the, concept of, 107
Perkin-Elmer, 57, 60, 61–63
Planets, determining size of, 131
Plato, 3
Pogo, 99
Popper, Karl, 126
Practical reason, structure of, 16
Pragmatism, 4, 40
Price, Derek, 123
Principia (Newton), 97
Project definition, 37
Proxmire, William, 110–11
Publishing, importance of, in science, 5
Putnam, 52

Quadrant, 130

Radiation, low-level, 74
Radical design, 37
RAND, 56
Rational decision, defined, 21
Rationalism, 3–4
Rationality, 19–23
  flawed theory of, 16–17
  nature of, 16
  role of, 123
  See also CPR (Commonsense Principle of Rationality)
Reagan, Ronald, 111
Real, definition of, 134
Realism, convergent, 5
Reality, constructivists on, 127
Reasoning
  context for, 18–19
  general pattern for, 18
Reduction, notion of, 134–35
Reductionism, Sicilian realism and, 135–36
Reflective null Corrector Washers, non-approved, 59
Reflective null test, vs. refractive null test, 61–62
Refraction, theory of, 93–94
Refractive null test, vs. reflective null test, 61–62
Reichenbach, 126
Relativity theory, 124
Religious values, 103
Restructuring, meaning of, 115
Revealed preferences, 17
Rifkin, Jeremy, 110
Rogers, G.F.C., 35–36
Rome, roads of, 5
Roosevelt, F.D., 89
Rudner, Richard, 82
Ryle, Gilbert, 35

Science
  epistemological relations between technology and, 1–2
  interdependence of, 92
  as knowledge producing process, 24
  philosophy of, 7–8
  as pursuit of knowledge, 3
  social dimension of, 8–9, 127, 128
  technological infrastructure of, 114, 122, 125–38
  as theory-bound, 47
Science, 60
"Science fiction mentality," 117
Scientific change
  Kuhn on nature of, 38
  theories of, 122–25
Scientific inquiry
  aim of, 34
  self-critical process of, 34
Scientific knowledge, 32–34
  evolution of, 29–30
  and technological infrastructure, 137
  as theory-bound, 33
  See also Design knowledge; Engineering knowledge; Knowledge; Technological knowledge
Second-order transformations, 13, 14
Sicilian realism, and reductionism, 135–36
Smith, Robert, 56, 63
Social constructivism, 4, 127
Social critics and criticism
  assessment of, 71
  and assessment of blame, 53
  and ideology, 71–72
  and new technology, 117–18
  place of, 27–28
  and technological change, 120
Social norms, function of, 18
Social process(es), science/technology as, 123
Social science laws, technological laws as, 43–45
Social security records, and privacy issue, 116

Sociohistorical concepts, 28
Solar system, discoveries re, 131
Space Science Institute, 58
Space Shuttle program, 57, 83
Space Telescope, The (Smith), 56, 73
Spherical aberration, HST's, 57, 58
Spinoza, Baruch, 3
Spitzer, Lyman, 56
"S/T," 26
Standing reserve, nature as, 70
Starry Messenger, The (Galileo), 95, 130
Statement of Initial Conditions, 42
Strategic Defense Initiative, 111
Strong Programme, 123, 127
Structure of Scientific Revolutions, The (Kuhn), 8, 127
"Studies in the Logic of Explanation" (Hempel/Oppenheim), 42
Success, guarantee of, 21
Supreme Court, 85, 119

Tacoma bridge disaster, 45, 46, 47
Technical cases, types of, 45
Technical explanation, 45–51
Technique, Ellul on, 87
Technocrats, 101
Technological change
  and human development, 120–21
  meaning of, 114–15
  resisting, 117
  as social issue, 114
  and value conflict, 113–14
Technological Change (Mesthene), 90
Technological determinism, 50
Technological infrastructure
  definition of, 129
  as key mechanism, 132
Technological knowledge, 28–32
  context and transferability of, 41
  See also Design knowledge; Engineering knowledge; Knowledge; Scientific knowledge
Technological laws, as social science laws, 43–45
Technology(ies)
  approaches to analyzing, 66
  as artifacts, 94
  autonomy of, 87
  as challenge to values, 101–2
  as complex of activities, 104
  defining, 9–12, 128
  de-ideolizing philosophy of, 84
  and dynamics of change, 97–99
  epistemological relations between science and, 1–2, 24
  essence of, 68–69
  Heidegger on definition of, 70
  as humanity at work, 11–12, 128
  as ideologically neutral, 82, 84
  and ideology, 70–82
  as input/output transformation process, 13
  instrumental conception of, 70
  as knowledge using process, 24
  Mesthene's definition of, 10, 36
  momentum of, 106–7
  and morality, 112–13
  philosophy and history of, 132
  process of, 90–91, 93
  second use of term, 12
  social dimension of, 51
  as threat to democracy, 100, 102, 107
  and values, 82–86
  values of, 103
"Technology as Knowledge" (Layton), 34
Technology of discovery, definition of, 128
Technology *simpliciter*, 12
"Technology—The Opiate of the Intellectuals" (McDermott), 90, 91
Telescope(s)
  Galileo and, 92, 93–96
  improvements in, 131
Theory(ies)
  criteria for choice of, 118–19
  epistemological relations between mechanism and, 6
  and instruments, 130
Three Mile Island (TMI), 111
  accident at, 46, 73–74
"Through the Looking Glass or News from Lake Mirror Image" (Layton), 34
Tools
  humanity's use of, 91
  as ideologically neutral, 82, 84
  as-mechanism-in-general, 10
Transformation(s)
  decisions as first-order, 13
  first order, 13
  second order, 13
True, vs. correct, Heidegger on, 68–69
Turing Test, 22
Tychonian mathematical astronomy, 95

U.S. *Congressional Record*, 57
Unintended consequences, 46, 116
  as basic feature, 49
Universality, 42–43
Universality claim, 32–33
Utilities, 17

Value change, problem of, 117
Value conflict, 112
  and technological change, 113–14
Value judgments, 70
Values, 17
  arguments re superiority of, 119
  changing of, 85–86
  conflict of, 102–3, 107–8
  kinds of, 103
  role of, 118
  technology and, 82–86
Value systems, and social change, 101
van Helden, 131
Vertical Radius Test, anomaly in, 62–63
Vincenti, Walter
  on design, 35–38, 54–55, 64
  reiterative model of, 51, 53, 60
Vote, value of, as *in principle* concept, 105

Whale and the Reactor, The (Winner), 72
What Engineers Know and How They Know It (Vincenti), 35
Winner, Langdon, 72–75, 80, 101
Work, defined, 30–31
World Wide Web, 116